This book looks into the question of why so many of today's children have poor posture and in some cases even posture injuries. There can be physical reasons for these posture problems, of course; on the other hand, poor posture is often the expression of a disturbed emotional development.

In this medical adviser, physical therapist Renate Zauner explains the connections between posture, movement, and the personality development of a child. She points out for the first time just how much strain is placed upon a child's ability to cope by his surroundings and by a life style that is not geared to his needs, and she demonstrates how these strains can lead to posture injuries. The author gives detailed descriptions of the physical causes and consequences of faulty postures as well as the clinical pictures of the posture injuries. In addition, she shows parents what can be done to ensure that their child will grow into the adult world straight and tall.

With the help of the gymnastic exercises for preschoolers and school children, parents can effectively train their children's endurance, dexterity and strength, thus offsetting the child's poor posture and ameliorating posture injuries in his or her later life.

Renate Zauner
was born in 1925 near Magdeburg and grew up in Königsberg, Karlsruhe and Göttingen. She studied medicine in Göttingen for ten semesters, then broke off her studies after her marriage and the birth of her two sons. In 1959 she took the state examination for physical therapy in Göttingen. Mrs. Zauner is the author of several medical books for laypeople, including *Speaking of: Backaches*. In 1961 Mrs. Zauner and her family moved to Munich, where she has her own physical therapy practice.

Helmuth Müller, M.D.
Professor of Medicine, was born in 1909 and studied medicine in Munich. His professional activities have included work in London and Munich. From 1947 to 1950 Dr. Müller was the head physician of the Children's Hospital in Bad Wiessee and he later worked at the Children's Clinic in Bethel/Bielefeld. His specializations include: research on the disposition to diseases, adolescent medicine, psycho-pathology of childhood. Since 1974 Dr. Müller has been active writing.

Speaking of:
Children's
Posture Problems
and the Injuries They Cause

by Renate Zauner

Introduction by Helmuth Müller, M.D.

Translated by Susan Ray, Ph.D.

Belair Publishing   new york

Library of Congress Catalog Card Number: 80-68764
ISBN: 0-8326-2244-3

Originally published in German under the title *Sprechstunde:
Kinder haltungsschaden,* copyright © 1978 by Grafe und Unzer
Verlag, Munchen.

Contents

Introduction:
A Word to Parents

Have you ever watched a small child stand on his tiptoes, stretch his back and arms full length and just manage to reach the doorknob with the tips of his fingers? The door opens a crack, and — success! The child turns around with a quick glance, and his face lights up with that mischievous smile that no one can resist.

What we see as perfect grace results from all those movements that together make the whole body the expressive organ of the soul: it can be the action of a young boy, a shy glance from below which signals a wish or desire, or it can be the bodily rigidity and the readiness for flight which express anxiety. Children have a particular ability to make themselves understood through body language; their hands and feet are still completely unself-conscious and extremely flexible, their backs are tall and straight, their necks hold their heads upright; and their heads seem to be constantly moving in every possible direction. One characteristic of primitive peoples is their ability to express themselves through body language, and this ability remains with them until long into adult years. How then is it possible that after only a few years in school, the eloquent body of our infants can turn into a child's body with poor or even faulty posture?

The child's body language

This medical handbook takes a critical look at the circumstances which lead to the widespread posture problems among our growing youth, and it will show the connections between emotional and physical posture. Life in our times has

7

undoubtedly changed and, although this has certainly expanded our view, it has also limited our physical freedom. We no longer walk barefoot on the forest floor; our children run on asphalt and cement and they have to learn how to sit still in school — simply because school offers no alternative. Up until a short time ago a languid or a "sloppy" posture was considered chic, but we are gradually beginning to again appreciate a normal, muscular figure. This change of attitude will hopefully also sharpen parents' attention to a beginning deterioration of their children's posture. As far as their physical growth is concerned, our children develop very quickly, they are tall and lanky, and thus more susceptible to posture injuries.

Tall children are more susceptible

As physicians, it is our duty to examine a child completely whenever he or she is brought to us, be it only for a cough, for this is the only way early symptoms of posture deviations can be recognized and treated in good time. Ours is a problem of differentiation: what is meant by the term "poor posture", and at what point can the problem be considered a posture injury? Is shortsightedness the reason why the child hangs his head, is it a case of shoulders that are fixed in a forward position, or are we dealing with round shoulders, or a lumbar flexure? Is the posture problem caused by foot and leg problems, by insufficient muscle tonus, by deformity, by "weak connective tissue"? An x-ray is usually not needed to determine these things, for more often than not x-ray shows less than what one can see by carefully observing the natural movements of the child while he is running, climbing, bending, leaning, jumping, and lying down. When the symptoms of a clear impediment can no longer be overlooked, it is already almost too late for successful treatment.

Timely recognition is important

I have worked for many years with physical therapists as co-workers, and it is from them that I learned the art of seeing disturbances in movement. It is always best for the examining physician to call in a physical therapist when he examines the patient, even when the ailment involves disturbances that emanate from the central nervous system, for these represent a large number of the disturbances children suffer. The author and the publisher of this book deserve great credit for explaining to the physician as well as to the parents of children with poor posture not only what the individual symptoms of incipient injuries are, but also the way they are usually treated. Moreover,

8

with the help of instructive illustrations and photographs, this handbook also shows what the individual can do. The road to "good posture" is long, to be sure, it does require endurance, and not only for our children, but for their parents as well, for they have to exercise their imagination in devising new programs for their children. This book provides many practical ideas for this very purpose.

As parents you can help your children develop a natural body feeling, to derive joy from movement, and to keep this sense of joy throughout their lives. As a result, your children will gain a sense of balanced self-confidence, strength and agility — characteristics which will "strengthen their backs" in their lives as adults.

Helmuth Müller, M.D.

1. The Child's Surroundings: Do They Meet His Needs?

Why Does It Have to Be My Child?

One of the most frequent questions parents of children with poor posture ask physicians and therapists today is what causes the visible flagging of their children's posture, why is this increasing deterioration of posture happening?

I, too, am repeatedly asked by parents: "Where does my child's poor posture come from? We do everything for his health, he doesn't lack anything."

As a matter of fact, when compared to children of earlier generations our children seemingly do not "lack anything"; they obviously know no renunciation, they do not not have to work hard, they usually have their own room to play in, frequently a small backyard as well, they are well-fed (sometimes almost too well-fed) and still more than 50% of our beginning school children have conspicuous postures, many of them even have posture injuries.

Rickets, the main cause of posture injuries in earlier generations, can no longer be cited as an explanation because, thanks to an intensive program of information and prevention, this disease has been brought under control. Earlier generations also knew children with weak connective tissue, and this remains the cause of poor posture in a number of today's children.

Something essential in our children's surroundings must therefore have changed, something that is having an unfavorable

50% of school-beginners have faulty posture

11

and deteriorating effect on their posture. Just where does this change lie, what is it that is making it more and more difficult for our children to grow up "straight and strong"?

In trying to discover the reasons for this, I should like to begin by examining the environment in which our children grow up with a view toward determining possible disturbance factors, After all, it is only when our children's living conditions coincide with their living needs that they have a chance for an optimal development.

Disproportion between Stress and Ability to Cope

More and more frequently pediatricians, orthopedists and psychologists are noting in studies the situation of the child in an increasingly child-hostile age. Children react in a very definite way to their surroundings, to the living conditions to which they are exposed. If these living conditions change, the child will also change.

Children often lead inappropriate lives

Today's generation of children is exposed to profound changes, and they must adjust and come to terms with them. In this adaptation process the constriction of their living space, the compulsion to adapt to unsuitable life patterns, and the lack of sufficient opportunities to gratify their strong motor (exercise) needs are particularly burdensome for children. Add to this the pressure that school, parents, immediate surroundings, and expectations of success exert on a child, and it becomes clear that for many of them there is a disproportion between stress and their ability to cope with it, in other words, between demand and the child's strength.

A child always reacts with his whole personality to the stimuli of his surroundings. He answers with his body and mind, with posture, movement, and emotion. This is why posture, movement, and emotion cannot be separated from one another when dealing with a child. The child expresses himself entirely through his body; feelings such as joy, tenderness, stubbornness, defense, fear, and oppression are mirrored in his posture, every mood becomes a gesture and a movement.

The main concern of this book, therefore, is not only posture injuries, but also the behavior patterns of a child. Orthopedic problems must be considered against the total picture of a child's personality.

In order to understand these connections, however, one must be familiar with the normal biological development of a child, and this includes all aspects that play a role in forming the child's posture: **statics** (body control, in the sense of balance and posture achieved by proper stressing of the joints), **motor function** (the complete movement process guided by the cerebral cortex), **emotional maturity,** and the **gradual development of intelligence**. For only after we know the norm can we recognize and evaluate the deviations from it.

2. Posture — Movement — Personality

The Child's "Motor Personality"

A child's posture is not a static process. Seen from a mechanical point of view, posture is the result of a muscular state of equilibrium in which the functional activities of the various muscles — the agonists and the antagonists — balance themselves out. Any imbalance in this interplay understandably has to disturb posture. If one now also realizes that posture necessarily has to be the starting point of every new movement, then the close connection between posture and movement becomes evident. Without the static element of posture, movement would not be possible at all. And the more stable the starting or basic posture is, the more perfect and correct will the movement that builds upon it turn out to be.

The connection between posture and movement

How well a child moves depends on the functional efficiency and the maturity level of his motor function on the motor command of the cerebral cortex, and on the movement itself. It is therefore dependent on the condition of the brain, of the nerve fibers and of the musculature that executes the movement. If the motor function is disturbed during the course of a movement, the movement, too, is disturbed. This disturbance can range from the minimum of being clumsy, awkward or stiff to the maximum of being uncoordinated, over-extended, uncontrolled or spastic.

Like other personality traits the motor function, too, develops in various ways depending on the individual. Thus one can speak of a child's "motor personality" or even of his myopsyche, of his "personality expressed via the muscles," or of his "muscle personality."

This individual motor personality already starts to develop in the first months of life. It then continues to develop and eventually becomes fixed in later years. It mainly consists of the tempo, dynamics and rhythm of a course of movement. Other authors, however, also include in this picture the child's strength, automatic associated movements, his ability to perform simultaneous movements and his motor coordination.

Relating these considerations to posture as the starting point of movement, it soon becomes clear that posture and movement are two components of one process, that they are mutually dependent, and that they can and do influence each other. More importantly, it also becomes evident that a correction of posture is very probably possible through training, whereas any single attempt at purely static exercises would fail to improve posture.

There is no doubt that good motor ability, and in connection with this, a balanced posture, is also a gift. Many children seem to be born with a natural grace and agility in their movements, while others are clumsy and awkward. But all children, not only the gifted ones, can be trained within a definite range to have good motor ability and posture. The motor process is stimulated by its surroundings. Objects that encourage movement, such as athletic equipment, and rhythmic instruments, such as drums and tambourines, direct and train a child's motor fitness. The child's need for imitation, too, as we shall see, can be directed and meaningfully applied to a motor activation. The natural stimulus for gross motor activities, such as running, jumping, hopping and climbing, however, should be present in the child's everyday surroundings. A large, child-oriented room, for example, should always be available, an area that encourages the child to move and play in such a way that he will make use of this opportunity as a matter of course and without being told to.

Support Tissue: the Muscular System

Just as a child's motor aptitude is partly inherited and partly acquired, the quality of his muscles, bones and ligaments can also be consciously improved, although they are genetically determined to a certain extent. Since the connective tissue is important for the mechanics of the whole support and posture apparatus, any weakness in this system will always have an effect on posture. Children with weak connective tissue are always children with poor posture; moveover, they are also usually lacking the supportive strength of a well-developed muscle system.

The supporting strength of the muscular system

One can improve the quality of tissue most efficiently through its function. The clearest example of this are the muscles: their primary function stimuli are tension and relaxation; they are improved by the easing and increasing of strength in measured doses.

It is more difficult to visualize this process being applied to the apparently rigid bones; yet for them too there are specific stimuli which trigger their growth and improve their quality. Here we should mention the alternation of stress and relaxation, the increase or decrease of pressure. Whereas stress or, in other words, pressure, works to inhibit growth, relaxation encourages bone growth, and this helps explain the increasing deviation of a knee joint toward knock-knees. In this condition, a considerably greater pressure due to flexion weighs on the outer half of the knee joint than upon the inner half, and thus the inner half of the joint is relaxed. The joint therefore grows stronger on the inner side than on the outside, and this leads to the leg position known as knock-knees.

The quality of muscles, ligaments, and tendons can also be improved by "firming" them, that is, by frequently walking barefoot, by brush massages, swimming, a lot of fresh air, and sun. Another important factor is a correct diet, especially as far as vitamins are concerned, for they play a great role in the growth and stability of bones.

Bone growth — bone stability

The musculature, which steadies weak joints and can raise and support the spine, provides essential assistance in the stabilizing of a weak support tissue. This is why a systematic training program concentrating on increasing the strength of the muscles and on improving posture is absolutely essential for children who consistently "droop", whose ankles cave inward, whose knees curve backward and overextend themselves, whose pelvis is tipped forward, whose shoulders droop or are drawn forward and result in protruding shoulder blades. Motor training, on the other hand, is not necessary here, because it is already present to a great degree.

Overweight Children Have a More Difficult Time

Overweight children, too, are often poorly endowed when it comes to their muscle systems, for they have weak tissues, little strength, and in many cases a strongly developed lymphatic system. These lymphatic glands together with large tonsils make overweight children more susceptible to diseases than their small and wiry peers, who are usually physically more resistant.

Fat children are also at a disadvantage with respect to fitness and endurance, because of the excess weight they have to move about. They not only lack strength, speed and physical agility, they are also less able to bear stress and tire quickly. Moreover, the fact that they are frequently exposed to the often cruel teasing of other children also hinders their development. By the time they reach kindergarten or the first grades of elementary school, they already sense their clumsiness, a slowness and early fatigue which make them shy away from physical activities. This in turn only makes them more apt candidates for posture problems. Many of these youngsters would profit greatly from a goal-oriented training program which includes a great deal of motor stimulation and posture correction.

Lack of strength

3. The Physical and Mental Development of the Child

Physical Growth

The growth of the body is one of the most visible signs of a child's progressing development. Of course, one can compare the height of a particular child only against the average height of his or her corresponding age group, and even this is not always valid, because there are various other physical maturation processes that have to be taken into consideration as well. Within a particular age group there are considerable variations in height.

The body does not grow in every direction at the same time; different organ systems have different rates of growth. There have been attempts to categorize these differing rates of relative growth. The four major categories that are usually emphasized include the growth of the nervous system; lymphoid growth, which refers to various glands such as the thymus gland and the lymph nodes; the growth of the bones and internal organs; and genital growth, which refers to sexual maturation.

Although all of these systems develop at a parallel rate during the prenatal, that is, the pre-birth period, their growth rates differ greatly after birth. The nervous system develops most quickly; in fact, it is already finished by the time the child is four years old. The lymph ducts reach maturity in a six-year-old, continue growing (even double in growth) during the next six years, and then decline. Basic genital development only starts with puberty.

The skeletal growth undergoes two main growth spurts, one
during infancy and one during puberty, and it ends only after
puberty.

Posture and movement depend essentially on both the
development of motor function, that is, upon the interplay of
brain, nerves and muscles, and on bone growth and the quality
of muscle and connective tissue.

Every organ system has its own specific growth stimulus
which, as we know, lies in its respective function. Movement is
the most important stimulus for motor function in the sense of the
motor command originating in the brain (the central nervous
impulse) as well as for the performing organ, the muscle. In other
words, movement is the most important function stimulus for the
whole neural-muscular system.

For the passive motor apparatus, on the other hand, for bones,
ligaments and joints, mechanical push and pull stimuli play the
major role. The main function stimulus here, then, is direct stress
and demand.

Bone tissue is at first very soft, for it is cartilaginous and
contains fluids. This bone tissue is transformed through deposits
of calcium salts into more solid, more brittle bone, and this
transformation process is fostered by a lot of fresh air, light, and
vitamins, mainly vitamin D. Vitamin C also plays an important
role in the formation of bones. States of vitamin deficiency during
this growth phase lead to an incomplete bone formation; the
bone remains pliable and elastic and thus becomes susceptible
to deformations.

Longitudinal growth of the long tubular bones of the limbs
takes place in especially formed growth zones called symphyses.
These zones separate the outer ends of such a bone (epiphyses)
from the shaft (diaphysis). These growth seams (epiphysis

symphyses) close at the end of puberty and thus permanently
finish a person's growth in height. Since the epiphysis
symphyses consist of extraordinarily sensitive cartilaginous
tissue, they are very easily injured. It is extremely important,
therefore, that they not be overtaxed, especially not
overextended, during infancy. This is why, when doing any
exercises with preschoolers, the supervising person must see to
it that the child's whole joint is included. This can be easily done
by supporting the lower part of the arm or leg in order to avoid

19

overly strong pulling effects on a particular joint.

The skeletal growth and the quality of the support tissue are dependent, but not wholly, upon inherited factors. The social environment also plays a role here. This explains the lesser skeletal height growth among children and adolescents who have to perform heavy physical work at an early age in comparison to students who have much less physical stress demands. Moreover, growth in height is usually great except during an extended period of illness. It is also not equally distributed over the twenty years during which the young person is growing; instead, it begins with a marked growth spurt during the first two years of life characterized by especially great progress during the first year, and then runs along relatively constantly until the second growth spurt during puberty. At this time the young body is once again particularly susceptible to posture deviations. Curvatures of the spine (scolioses), round shoulders or hunchbacks, which were perhaps only slightly apparent during the early years, can now become more obvious.

Disproportion between skeletal and organ growth

The internal organs, too, mature at different rates. Thus it is possible, for example, that there can be a temporary disproportion between the skeletal growth (that is, height) on the one hand and the growth of the heart and the circulatory system on the other. This in turn can result in a lowered ability to bear physical stress. This fact has to be taken into consideration during all gymnastic training programs. It is precisely the tall and lanky adolescents, the ones who have grown up very quickly, who frequently experience real states of exhaustion brought about by these kinds of varying growth spurts. These conditions are frequently interpreted, unfairly, as laziness on the part of the young person. In an attempt to cure an apparent malady, these young people are then required to participate in all kinds of sports activities which at this point in time actually overtax them. In these cases it would be best to let a physician decide whether or not one should perhaps wait for a later maturation of the heart and the circulatory system before undertaking any intensive training program.

Motor Function as Motor Impulse

The phrase *motor function* refers to the total neuro-muscular process ranging from the motor command to the execution of the movement: in other words, it involves the presence of the central nerve cells (located in the brain), which give the impulse, the presence of the peripheral nerve which transmits the stimulus, and the presence of the muscle, which executes the movement. Nevertheless, this in itself would not be enough to enable a person to move in many different ways, to develop agility, and to coordinate and distribute movements. For this the sense organs are also necessary, the senses of seeing, hearing, feeling, and orientation in space. This interplay of brain and nerves, muscles and sense organs is usually grouped together under the term *sensory motor function.*

Sensory motor function

Just how very dependent we are in our functional ability upon an undisturbed course of motor function becomes clear as soon as one member of this functional unit fails to operate properly. A central disturbance in the nerve cells of the brain, for example, alters the basal tension of the musculature which is regulated by this central nervous portion of the brain; tension and relaxation of the muscles are then no longer distributed in correct doses. Other disturbances can find their expression in coordination difficulties.

Finally, an injury to the conducting nerve prevents the motor command from reaching the muscle with the result that the latter does not move at all. The main source of motor function located in the brain can be considered the more important part, since it constantly embraces and influences the entire body, whereas the effect of paralysis of one single nerve can be confined to a small part of the body. Since motor function co-determines all of a person's smallest unconscious movements, it also shares the responsibility for motor expression, for mime and pantomime — those things that make an individual recognizable even from afar. These traits include a person's manner of walking, of turning his head or of reacting to fright or joy. Motor expression and basic emotional mood are thus closely related to one another, and this is why individual motor function is also always a mirror of a person's temperamental disposition at any given time. Naturally,

Much depends upon motor function

21

this is particularly true for children.

Motor function, however, in addition to possessing this expressiveness, also indicates the maturity level of the child's brain. The child's neuromuscular functional ability and sense of coordination also mature to the same extent that the cortex and the underlying subcortical portions of the brain form and develop their centers and tracks. The fact that a three-year-old child, for instance, is not yet able to hop on one leg is not due to a still untrained muscle system, but to a still immature neuromuscular system, to a lack of balance and an inability to "comprehend" this difficult movement. And if an eight-year-old child whose total development is retarded cannot yet ride a bicycle, it is not because he is afraid, but rather because his sense of coordination has not yet matured far enough to be ready to ride a bicycle.

Many ostensibly physically clumsy children have a slight neuromotor retardation which makes them simply awkward when it comes to physical movements. They jump stiffly and hold their arms and hands in a cramped position when jumping down from a chair or when jumping on a trampoline without demonstrating any other disturbances. Many of these difficulties disappear by themselves in later years because the organism can learn motor patterns and sequences to a certain extent. Even more conspicuous motor disturbances can be obviated if they are treated by a goal-oriented program of exercise under the guidance of a physical therapist.

Motor distur-
bances can
be rectified

"Borderline cases" of disturbed motor function in a child, which are far more numerous than once assumed, thus require extraordinary attention on the part of the parents. These children need a great deal of motor stimulation, guidance and patience; most importantly, they must be watched very carefully to ensure that their movements and posture develop normally. A well-known child psychiatrist, Dr. Hünnekens, once compared an infant's sensory and motor canals to unused fire hoses, whose inner surfaces were still resting against each other. They require the pressure of the environmental stimuli to gradually inflate them. Through the stimulus of their surroundings they become open. In fact, various studies have been made to demonstrate just how significant the external stimulus is for the development of motor function. Studies of this kind have shown that the motor

Children
need external
stimuli

22

development of a child in our culture clearly slows down around the age of three, whereas among children of many primitive races it essentially improves from this age onwards, especially in so far as agility and the ability to respond to one's surroundings and one's environment are concerned. Obviously, during this developmental period our children lack the necessary challenge to awaken their activity, their desire to play and their sense of adventure, as well as the opportunity to have physical motor experiences. Underlining the truth of this statement is the still predominantly stereotypical and sterile design of today's playgrounds with their uniformity of equipment and paved schoolyards, which are good for very little else than scratching and cutting one's knees. Today's children's rooms also deserve mention here, for as a result of the basic concept of apartment planning and furnishing, they are destined to be too small. Because of this they set much too narrow constrictions on children's play, which requires active, physical movement.

The Child's Ability to "Comprehend" His Surroundings

What are cognitive functions?

This section, which deals with cognitive functions, may at first seem outside the scope of this book. In fact, however, the collecting of experience, the "comprehending" of one's surroundings in its literal sense, depends on the smooth working of motor function. The term "cognitive functions" refers to the processes that transmit "information." This means abilities, such as perception, imagination, recognition, the proportionate relationship of things. The functioning of our senses of seeing, hearing and touch is also a part of this. It therefore has to do with an area in which an individual's motor achievements, sensory fitness and perception all work together. And for this area of a child's development, learning processes and stimulus — a toy, for example, that helps children "comprehend" their surroundings and collect experiences — also play a very decisive role. Blocks that can be arranged, that can be nested within one another, can be built into a tower, objects that can be taken apart and examined are all things that any child could hope for; to put in his mouth, to bite, to smell, to feel, hear and see. Tools that foster motor dexterity, as for example hand tools, also help indirectly to expand a child's knowledge of things.

Children and street traffic

The example of a child in street traffic clearly demonstrates exactly how vitally important the cognitive maturation is for a child. It also shows just what demands are already made at a very early age on the functioning of his cognitive abilities of seeing, hearing, understanding, evaluating and acting. This necessity to concentrate on everything at once overtaxes children a great deal because their cognitive development has simply not yet progressed this far.

It is no wonder that so many children are injured or killed every year due to traffic accidents. Children are often the victims of grown-ups who do not accept, or are unwillng to accept, the fact that children are not adults and have not reached adult development.

Another fact in this context is a child's marked drive to imitate, which is his means of learning during the first decisive years of life. What adults offer him as learning models in street traffic — disregard for rules and careless behavior — hardly needs detailed discussion here.

Connection between Mind, Posture, and Motion

Movement is a basic need

A green field sporting a group of romping children — one can hardly find a more cheerful, more dynamic sight than that of a group of children charging out of the confines of kindergarten and into the open fresh air. With what unconscious charm, grace and vitality these young people move about! Hardly any child of this age "walks" — movement is above all else recreation, it is a basic need that has to be satisfied.

This dynamic scene is less frequently seen in the school yards of the upper grades. Many aggressive components are now mixed in with the light-hearted fun; fatigue and disappointment enter the picture and are reflected in a child's movements and posture. Here shoulders and head hang if a child is criticized for poor schoolwork, whereas a child hops about when he is exuberant and happy. In the schoolyard it is very easy to distinguish between the successful students and the "failures".

There are many common sayings that aptly express this close connection between posture, movement and personality: we say that a person has no backbone, that he loses his bearings, that someone has to straighten him out. A person can be bowed down by grief, can have an upright character; we can jump for joy, or brace ourselves for trouble.

Disturbances caused by continuing setbacks

This association of mind, posture and movement is more strongly marked among children than among adults, because a child expresses himself with his whole body via his pronounced motor function. This association also shows just how important the physical and the emotional ground work is for the total overall personality development of a child. A child can be externally and internally "bent and disturbed" his whole life long if he has to repeatedly experience failures in the presence of his peers. On the other hand, a child can also flourish and prosper when his self-confidence and his "backbone" are strengthened as a result of an increase in his physical strength and fitness.

From a Crawler to a Toddler

We have already established the fact that posture is the expression of the total personality. What now remains is to consider posture from an orthopedic viewpoint.

One of the most essential and observable stages in the development of a child's posture is the transition from going about on all fours to an upright position, the transition from a crawler to a toddler. When this transition does not occur harmoniously and symmetrically in all parts of the body, or if it cannot do so, subsequent posture problems dating back to this period are bound to develop.

Position of the pelvis

This important development from the crawling to the walking child involves two main problems: one is the upright position of the pelvis, which is acquired only over a longer period of time, and the other is maintaining balance, or the ability to steer one's motor function.

In an infant as well as in a crawler the hip joints are still strongly bowed and their legs only seem to stretch straight while the child is lying on his back. What is actually happening here is that the extension of the hip is attained only via a lumbar flexure with which the child stimulates the lacking hip extension. The child is also compelled to form this lumbar flexure if he wants to stand upright; what he lacks in the stretching ability of the hip joint is replaced by a pelvis tipped forward and a correspondingly protruding stomach. A very flat back then rises above this tipped pelvis, there is hardly any neck to speak of, and the head is prominent.

At this age a child's torso movements rotate for the most part around the motor axis which passes through the hip joint. This means that, when a child becomes unsteady on his legs, he simply bends them and sits down. Fortunately, given the shortness of the legs, this is a painless process. In order to get back on his feet, the child supports himself on both hands, spreads his legs and feet wide apart and stands up, still wobbly and still trying to keep his balance.

Up until about the second year, a child's gait remains quite spread out with short steps. Only toward the end of the third year, when the child has extended himself to his full height, do the hip flexure and the marked lumbar flexure disappear, the neck becomes longer, and the head smaller in relation to the rest of the body. The body now learns how to balance itself more securely and not only how to walk, but how to run as well.

At this age the first signs of weak connective tissue will begin

First signs
of later
problems

to appear among those children who are prone to it: this weakness appears in the form of talipes valgus, flat feet, and an increasing tendency toward knock-knees. A hip kyphosis (curvature of the hip) will also reveal itself during this toddler age, especially among those children who have learned how to sit at an early age and who continue to sit a great deal, while deviations in the upper regions of the spine become visible for the most part only at a later date.

The natural curvatures of the spine are already clearly visible in the school child in the slightly concave contour of the small of the back, the rounding in the chest area, and the slight lordosis (forward curvature) in the cervical vertebra.

Posture
habits

Many children in this age group develop the habit of making a lumbar flexure with its corresponding protruding stomach. This is a "relic" from the time of the early upright position of the torso. It is particularly important precisely at this young age to transmit to children a sense of posture which will make them aware of the position of their pelvic girdle. A faulty position of the pelvis forms the basis for a large percentage of poor postures and later lower back problems, a common orthopedic problem for adults.

What does "good" posture look like? Perhaps the simplest description is a functional one: a *good* posture should also always be a *correct* posture, which is to say that the person's back tires the least with a well-balanced posture. This is because

The balanced
spine

a spine curved at the right places in physiological distribution rests in itself. It neutralizes the convex and concave curvatures, it is well-balanced, maintains its own equilibrium, and bears its own weight. Every imbalance, every deviation, regardless of the direction, disturbs this framework and requires extra work of the muscles and the support system in order to compensate for the lacking equilibrium, and this is tiring.

This complete bodily equilibrium and symmetry is related to the skeleton, the position of the joints, the position of the pelvis; it is also related to the uniform development and the strength of the muscles and a functional balance within the nervous system. Every disturbance of these mutually dependent systems results in a deviation of the total statics and this becomes visible in poor posture. It constantly demands additional strength from the child, since the natural equilibrium with which the body balances itself is disturbed. This child will therefore fatigue more easily, will "slouch," and will neglect his posture. The goal of the following suggested posture exercise program lies primarily in returning its balanced supporting strength to the back so that the child's strength will suffice to absorb disparities and maintain the extension of the back.

From an Infant to a School Child

In view of its place in the world, an infant still has it relatively easy. He is the uncontested focal point of the family. He soon learns that his helplessness provokes the undivided attention and assistance of grown-ups, for he only needs to scream his discomfort to the world and help is already on the way. He will be fed, picked up, his diapers will be changed. The infant thus builds up for himself a strong "ego role" which lets him feel almost omnipotent. As he grows older, however, he soon discovers that the rug is increasingly being pulled out from under his feet, his demands are systematically restricted, and his parents expect certain considerations from him. Their attention will be turning once again to their own lives, and they will expect an increasing integration of the child into the rhythm of their adult lives.

The infant's ego role

At about two and a half years, the child gradually begins to realize that his position in life is by no means as all-embracing as he had at first believed. He senses that he is for the most part weaker than he believed himself to be, thanks to the constant assistance of adults, and he is gradually introduced to the new experience of criticism. A child does not process these new experiences without certain inner difficulties which are expressed in defiance, obstinacy, and aggression. In view of this situation the child has really only once choice: he abandons his magnificent status and takes sides with what in his eyes is the next most powerful authority in his surroundings — and that is almost always his parents. This "hanging on" to his parents is so complete that one can almost describe it as a "satellite formation" between the child and his parents. This phase lasts in this state of total exclusion until about the time the child is four years old.

The child's readiness to acknowledge his parents' omnipotence is, of course, repeatedly interspersed with strong expressions of selfishness; these are a carry-over from the previous period of autocracy. The child's needs in this period are still very unsocial, for he enjoys playing by himself and cannot do much more with playmates of his own age than he can with a toy. On the other hand, he eagerly adopts all sorts of skills and

The infant's needs

abilities from his parents and is extraordinarily receptive to their praise. His preferable mode of learning is imitation. To his mind, the most wonderful things are those that a child does with his beloved parents. In addition, he has a great deal of imagination and identifies completely with each respective role that is presented to him. He takes situations that he has thought up himself very seriously, is curious, and increasingly creative.

This strongly dependent parent-child relationship gradually but inexorably dissolves when the child enters school, for now other people, teachers, for example, or sometimes a group of peers, are included in the child's love and admiration. The strong parent-child bond also suffers more and more frequent incursions due to the fact that, on the one hand, the child begins to experience his own progress and the possibility of a certain independence more and more consciously. On the other hand, the ever more frequent and ever more necessary parental prohibitions plainly show him his limits. The child perceives his own impotence with regard to the world of "grown-ups". It goes without saying that this process of conforming and falling into line is not without its rough spots and friction, and thus there are repeated outbursts of rage and aggression during these developmental years.

The child's powerlessness with respect to the adult world

The child's attitude toward play is also constantly changing. Whereas a four-year-old still predominantly plays without regard to his peers, the five- to six-year-old loses a bit of egocentricity and becomes increasingly more interested in other children, whom he now begins to differentiate and to assess. Friendships start to form. There is an increasing readiness for self-exertion and accomplishment; the child becomes increasingly independent and self-reliant, and he becomes ready for school.

We will return to these various psychological behavior patterns in the preschooler and the school child when we discuss the choice and organization of the exercise program. An understanding of the egocentricity and the elevated self-esteem of our little ones as well as an awareness of the incipient spirit of involvement and of an ability to compete in the school child can help us encourage our children to collaborate in a spirit of positive cooperation.

The Correlation between Age and Ability

There is always a certain risk involved in specifying dates as orientation points to help evaluate a particular child's stage of development. The temptation to measure one's own child against a norm and to adapt him to it as completely as possible is always great. Even if parents are objective enough to realize that the margin in which the developmental stages take place is very large, that it can even be a matter of five or more months at a time, there may still remain a small amount of ambition on their part. For the fact that it should take, of all people, one's own child a longer period of time in one area that it does other children of the same age group is often difficult to accept. The danger of forcing one's own child to follow a preconceived concept is great whatever form it may take. Nevertheless, there is a need for a certain orientation guide to alert parents to possible cases of actual developmental retardation and to encourage them to consult a physician. The following summary seems to be sufficiently encompassing and still suitable as a standard of comparison since it leaves a real margin for individual developmental differences. The developmental stages according to doctors Hofmeier, Müller and Schwidder (*Alles Über Dein Kind, Rororo-Taschenbuch, 1971*) look like this:

Orientation aids for parents

• **In the first year** the child should be able to raise himself to a sitting position and sit upright without support; to stand up with assistance and to open a box by himself; he should have learned the first language skills, such as the speaking of simple syllables and words like mama and papa.

• **In the second year** the child should be able to walk, run, and climb up on a chair, pile wooden blocks on top of each other, and name simple objects. He should also be able to fetch objects that are out of his reach by using a rod or a stick through the bars of his playpen; he should recognize pictures and be able to point out and name dogs, people, etc. He should also be using connected words and singing simple melodies.

- **In the third year** the body movements should be completely controlled, including jumping, climbing, and independent eating with a spoon. It should be possible for the child to build something simple with blocks, and he should begin to develop an almost complete understanding of words including the ability to repeat the new and unfamiliar ones.

- **In the fourth year** the child should show a balanced control of his body and be able to use simple tools; the ability to carry a full glass of water, the first beginnings of adapting to the playing rules of games, and the ability to determine by touch with closed eyes such objects as a bottle, a button, or a doll, all start here.

- **In the fifth year** the child should be able to correctly compare weights of different things, identify objects according to their use, carry out three tasks that have been assigned at the same time. The child should also be able to name the primary colors, solve simple patience games (such as guiding a ball through a maze), and find four out of five hidden objects.

- **In the sixth year** the child should be able to build a complicated structure that someone has demonstrated to him, solve a complicated game of patience, and differentiate between left and right, morning and afternoon. He should be able to name the days of the week, count the number of his fingers, and recognize four of the most common coins as well as the various components of a picture.

4. Factors That Can Disturb Development

What has been said up until now about a child's development was intended to assist parents in becoming better able to evaluate their child as far as his emotional and physical development are concerned, and to recognize his physical and emotional deportment. Yet, if a person really wants to help his child, he must also understand the possible disturbing factors that can inhibit or misdirect normal development.

How you can help your child

Such disturbing influences can be of an organic origin as well as the results of adverse living conditions. Parents can possibly contribute to these adverse conditions out of a lack of knowledge or ignorance. For even slight shortcomings in the living space designed by parents can add up for a child, they can limit his scope of development and thus have an adverse effect on his health.

Posture Deviations Due to Physical Disturbances and Injuries

There are a number of physical illnesses, changes, or malformations which affect and disturb a child's posture. If this book were to consider all of them, it would soon grow into a small medical compendium of chidren's diseases and as such would

neglect its actual purpose, which is to recognize posture
deviations in children as early as possible and to treat them with
a goal-oriented exercise program. This is why I have not
included malformations of bones such as missing collar bones or
redundant vertebrae, for instance, nor injuries, tumors, or
infectious diseases of whatever type in the following discussion.
Complications due to brain damage have also been omitted,
since they need the treatment of specialists in the field. Clubfoot,
a child's wryneck, which can be congenital or acquired, growth
disturbances caused by hormonal or infectious changes, as well
as Scheuermann's disease will not be discussed here.

There are, however, some health factors that are important
for a child's posture, and near-sightedness is one example. It
encourages a child to develop a writing posture with a rounded
back and can even make his movements unsteady. Poor hearing
in one ear is another factor; it can cause a child to unconsciously
slant his head to one side; one leg that has become shorter than
the other due to an injury can twist his spine. The point is that
every bodily asymmetry, be it ever so slight, brings the body out
of its finely balanced equilibrium and disturbs its posture.

The Wrong Walker

Following all that has been said up until now about the
motor needs of a small child in connection with his healthy
development, it would almost seem obvious that keeping a child
cooped up for hours on end in the modern walkers made for
toddlers is not conducive to his development, neither for his
enormous motor drive, for which a walker is a hindrance, nor for
the training of his gross motor activity, for which it is actually
injurious. These walkers, with their woven net walls, just about
force the child into a static faulty posture; they only allow sitting
or standing, and at best a hanging position. The name "walker"
is hardly justified for this oversized "butterfly net". It accustoms
the child to an excessively long period of sitting because he can
lean comfortably against the much too compliant wall which then
buckles out in a rounded shape conforming to the contour of his

back. Moreover, there is no room for a comfortable restful abdominal position because of the many playthings attached to it, and the narrow space and the thick netting that hinder his sight inevitably encourage the child to work his way up to a standing position in order to be able to see freely and grasp things over the edge. Even the fetching of an object the child has thrown out beyond the limit of the walker — this important learning process of "doing, grasping and comprehending" — is missing here. And so little overtired children "hang" on the edge of their walker and hold themselves iron-tight until their eyes close and they fall over with fatique.

Surely many hip kyphoses in preschoolers are the results of staying in such a small walker for too long at a time. There is also no doubt that a retarded start of the crawling phase is associated with the lack of any training opportunity and the lack of employment of the gross motor activity. The result of all of this is that the child's back remains weak because there is no motivation for lying on his stomach, his back tires prematurely because he has become accustomed to sitting for long periods, and it becomes rounded because of the lack of support in the flexible walls of the walker.

No incentive to crawl

Ideally, this kind of walker should be used only for a short period of time, not as a daily playpen to which the child is abandoned for hours. There is an alternative possibility in a wooden walker with rungs that can be bought, or even better, made in generous proportions so that it measures at least six feet (two meters) by six feet. At the beginning of the crawling phase one can also make a lattice gate that fits into the door frame of the child's room so that the whole room itself takes on the function of a walker or playpen. When the door is open, the child can keep a vocal contact with his mother, frequently even a visual contact, while at the same time he can play in a large expanse of space without getting into other rooms where there might be some danger. Children who are so extensively constricted in their activities by a small walker or playpen during this, one of the most important growth periods, are naturally unable to develop strong support tissues, and this handicap considerably increases their propensity for later posture injuries.

The child's room as playpen

Lack of Time and Attention

The need for attention

Part of what makes a child thrive physically is his parents' reaction to his activities and their encouragement of his abilities. By observing as well as protectively guiding the child, the parents also constantly stimulate his fantasy and his creativity.

Children have a great need for repetition, for common processing of that which has been seen and heard. Their thousands of questions a day, which can almost drive a parent berserk, are frequently asked simply so that the child's contact with adults is not lost and that he does not lose their attention. Small children, of course, prefer to play by themselves, but always with the possibility of a frequent contact with their mother. They like to demonstrate newly acquired bodily skills and tricks, and for this, too, they need their parents' response.

A properly administered direction of the development of physical strength and agility can be gained by performing small chores at home: by pulling over a chair and climbing up on it in order to fetch something new, watering the garden, pulling or pushing a wheelbarrow, carrying, shovelling and carting off grass or soil — all of these things require and contribute to physical strength and experience.

Children learn by experience

The fact that thorns prick and nettles sting is something a child will best learn for himself, for warnings do not help at all in these situations. And the fact that a shopping bag is too heavy to carry, or a desired object is hanging out of reach, is also something that everyone has to determine for himself before he thinks about asking for help. Disappointing discoveries such as these of one's own insufficiency are bearable only if they are made under the comforting encouragement of parents and are mitigated by their presence. If children are left to their own devices in all of these situations, if they lose their curiosity, their courage and their joy of discovery, then their agility and their strength will become stunted as well.

The Overburdened School Child

Sitting still is the greatest strain an elementary school child is required to bear. When considered from a posture viewpoint, it is almost tragic that no one has yet devised a form of learning that does not make sitting still at a desk a necessity. If one were to take the physical needs of this age group into consideration, these children would be provided with standing desks, be allowed active exercise at twenty-minute intervals, and then be allowed to lie on their stomachs to rest. Ironically, today's gymnastic instruction is another detrimental aspect, for it carries a compulsion to succeed along with it in the form of grades. This robs the recess period of one of its essential components, namely: the pleasure principle, the unchecked playful counterbalance to intellectual concentration. Grading deprives the physically less skillful children of the refreshing effect of this so very necessary physical relaxation. The inhibited even attempt, if possible, to withdraw, and the trend frequently becomes to skip precisely those very subjects that are fun to do, that develop physical strength and creativity, that provide stimulation and a chance to exert oneself positively. What the child is actually skipping, and thus depriving himself of, are those subjects that are supposed to foster the total personality, namely: gymnastics, drawing, and music. In our achievement-oriented society these subjects have been and continue to be woefully underrated. And if a great deal of effort is being made today to replace the telling label of "minor subjects" for these fields, it is being spent, at least on the secondary school level, only in order to categorize them as "achievement subjects". And that speaks for itself.

Another cause for concern is the fact that the physical and emotional preparation of our beginning school chidren is considerably less favorable than it was for earlier generations. Our children frequently do not lead a life that is appropriate for a child, but rather one that is appropriate for small adults. This already begins with the daily life rhythm. Not only the everyday routine, but also that of weekends and of vacations is not geared to a child and his needs. The fact that children frequently have only one afternoon nap and otherwise work through the day is already injurious, but the fact that in many cases they are also robbed of the very necessary walk to and from school because they are driven both ways in a car, is terrible. The walk to school has a real and highly essential function for the physical development and the maturing of a child's personality: it trains his

Exercising without the pleasure principle

Fostering the total personality

The child's daily rhythm

physical strength, attention, motor function, endurance, and it creates his first social contacts. To make friendships, to learn to defend oneself, to become used to wind and weather and to learn proper respect for street traffic, to be responsible for one's self without the control of the parental eye — where is the child supposed to learn all of this if not on the way to school? Most young mothers, whose cars line up in front of the schools, cite as reasons for their chauffering the children's heavy schoolbags and the risk of a traffic accident. The first problem can be alleviated if school books were carried in rucksacks on the child's back, if the straps of these rucksacks were correctly adjusted. If they hold the bag over the shoulder blades, the child's back will extend, but if they hang on the buttocks, they pull the child's back into a lumbar flexure. Another good idea would be to go through the contents of the school bags daily and remove whatever is not essential. Children frequently drag superfluous things back and forth simply out of habit.

Parental apprehension about the dangers of street traffic is undoubtedly justified. However, these dangers are not only present on the way to school, they accompany a child constantly — whether he or she is climbing out of the school bus, riding a bicycle or simply playing outside. A successful Scandinavian experiment has shown that this risk can actually be reduced if parents and children participate in a common traffic safety training program. This is one way to provide a child with protection against street traffic.

Not Enough Time for Play

Weekend plans in our society are also not geared to the needs of children. Here, too, exercise gets the short end of the stick. Apartments hardly allow for any running around. Restricted by space, frequently in one of the smaller rooms, and by the amount of noise they can make without bothering the neighbors, children have hardly any alternative other than to watch television. And in doing so they drape themselves over chairs or on the sofa instead of at least lying on their stomachs or sitting crosslegged on the floor. If they do go outside, they have the all too familiar playground which has no other challenges to offer than the unimaginative seesaws and monkey bars. Even walks are not designed with chidren in mind. Everything that is measured, uniform and geared toward endurance runs counter to a child's motor drive. And his curiosity, his thirst for knowledge, his need for dynamic movement, his need to acquire strength, agility and speed as well as to put them to use, are also not satisified. An example of what is geared to the child and his needs is camping in the open air, where he can climb, make discoveries, roughhouse and run about, or to have as large a room as possible at home as a play and recreation room. This can be a basement or attic room, actually any room in the apartment so long as it fulfills certain requirements. It has to be large and it must be variable, that is, changeable and shapeable. It may not contain any dangerous (pointed or angular) or fragile objects that have to be protected. Welcome are a rug on the floor, a large number of cushions, a wall painted in a cheerful color and one that the child himself can also paint on. Other important objects include gymnastic rings that hang from the ceiling, a rope ladder or a climbing rope, a hammock and materials such as large cardboard cartons, pieces of fabric, paper — anything that can stimulate a creative imagination. Furthermore, it would be more bearable for the nerves of the whole family if the children were given a large room of their own. This will help ensure against aggressive and physically underactive, nervous children.

If our children lack motor stimuli and opportunites for running and playing, they will also lack the necessary circulatory stimulation which is in the end responsible for their physical development, for their respiratory volume and for their cardiac fitness. A weak constitution will thus deteriorate further, and a child whose posture is in jeopardy can become prone not only to posture injuries, but also to diseases of the circulatory system and metabolism during his years of development.

Walks with children in mind

The largest room for the children

Exercise as stimulation for the circula- tory system

Television-Induced Disturbances in Development

Television exerts considerable control over the lives of our children. We have already drawn attention to the particularly languid, inactive, orthopedically unsound posture of most television watchers. But this is only one part of the problem. Paired with the external passivity (which is totally contrary to the needs of a child's organism) while watching television is an unphysiological activation of the child's autonomous nervous system — in other words, his emotions. Events on the screen affect a child much more deeply than they do an adult. A child cannot evaluate all this material that is bombarding his senses quickly enough, nor can he differentiate the reality from the make-believe. If he knew, as adults do, that it is all only appearance and game, he would not have to cry and not have to laugh out loud during the course of a relatively banal action. The child is physically checked by the medium of the television but he must emotionally complete what for him are maximum activities. Afterwards he is usually unable to work off these stimuli because no one explains them to him or gives him a chance to act out what he has just seen. Instead of this he goes off to do his homework or he goes to bed, and his parents wonder why he is not concentrating on his work or cannot sleep at night. What is the child to do with his emotional projections?

There is a real and measurable "action metabolism" which takes place when a person completes a physical act of maxiumum effort. It can be demonstrated and read from certain waste products of the organic and muscle metabolism in the blood. This emergency state of the body disintegrates naturally through physical activity. This same "action metabolism" is also put in motion during a passive experience, but with the important difference that it is not worked off by physically acting it out. From this moment on it becomes injurious, for it makes the individual aggressive and irritable.

This reaction of the organism does not only occur while watching television; this tension metabolism is also produced while riding in a car. In this situation, too, by far much more is demanded of the child than is proportionate to his means;

No opportunity to process experience

furthermore, it is again passive, and the child is forced into an adult rhythm. Many fathers point with great pride to the fact that they drove the thousand miles to a vacation spot without considering that the four- and five-year-old child had to do it, too, and could not work off his energy.

Lengthy automobile rides are not good

Frequent geographical changes, long automobile drives, penned-up impressions of constantly changing surroundings and being together with many strange people are extraordinarily great emotional and physical stresses for a child. Children need child-oriented proportions in their experiences, in the surroundings in which they live and in the undertakings to which they are exposed; if they do not have them, they suffer.

The Overprotected Child

Insecure parents — anxious children

Parents who are too anxious or worried about their child also inhibit his or her development. Insecure, anxious parents, unable to cope with the necessary adaptation problems in their own lives, lessen their own anxiety by shifting their worries to their child, especially since they can ostensibly better deal with a child's problems.

This, however, is not without consequences for the little one. He is not only kept dependent on others for a much longer time than need be and hindered from having his own experiences, but the anxieties of his parents can also be transferred to the child so that he himself becomes insecure and anxious. A child who is not permitted to take any risks will not learn to recognize his limits and will not learn to judge his abilities. He will not be able to defend himself because every attack, every threat is intercepted by his over-attentive parents. The child experiences this strong bond to parental protection as oppressing, and all the more so because he cannot escape from it. Thus the child goes back and forth between insecurity and aggressiveness. He frequently takes shelter in laziness and fatalism and passes all responsibility on to his parents: dressing in keeping with the weather, checking the school bag, determining the time he leaves for school in the morning. In these cases all of these things are taken over by the parents or are controlled by them. In addition, the child is also physically at a disadvantage, because there are many things he may not do: he may not climb trees, jump in the water, climb over fences, balance, swing, and run fast. Many attempts at expanding the limits of his own abilities may be blocked by his parents.

Wrong understanding of care

Since children suffer under this kind of dependence and this wrong conception of care, parents ought to consciously and repeatedly check and expand the child's opportunities to grow and to have his own experiences. Only in this way can parents adjust to the child's increasing abilities so that he may both learn to act on his own initiative, and learn a sense of proportion, self-assessment, courage, and an individual sense of responsibility.

Today's children have a much more difficult time in many respects in finding their place in the family and in society than earlier generations did, because for earlier generations the family was still the essential "school for life": it was the place where children grew into duties and responsibilities on their own, where there were workshops, gardens, a large household with many siblings and thus meaningful and real duties and chores through

Children mature through chores

which the children assured their place in the family. And it was in the family that they developed a sense of being needed, where they could "learn" to gather experiences and, most importantly, where they could also afford to make mistakes. In our adult world children are essentially shut off from genuine responsibility, the routine chores and small jobs to which they are raised are basically unimportant and can be done without. These responsibilities are not significant enough that a child can really mature with and through them. And so today's children look for their opportunities for self-realization outside the family; they prematurely turn to peer groups, for here they are not excluded from responsibilities, but instead are taken seriously.

Obviously, a common family-oriented program of gymnastic exercises will not obviate all these problems. Nevertheless, this kind of program does offer the possibility of including the older siblings in a kind of shared responsibility for the smaller ones in that they are made aware of the importance of this conscious physical education.

5. Posture Problems — Posture Injuries

Weak Connective Tissue and Rickets

There is a close connection between weak connective tissue and rickets, and for this reason I should like to discuss both of them in more detail. Severe disturbances of bone growth due to rickets are rare today because of effective prevention through early doses of vitamin D. Nevertheless, many children with posture problems are borderline cases of this disease, and their slight deficiency of vitamin D strengthens an already congenital weakness of the muscle and connective tissues.

Resorption disturbances

Rickets is not always an indication of a deficiency of vitamin D; it can also be a case of absorption disturbances, that is, of a disturbed ability of the intestines or the kidneys to absorb phosphorus and calcium. As a result of insufficient calcification, the bone remains soft and this can result in the curvature of strained bones (bowlegs, for example).

Since rickets also always affects the quality of the muscle system and of the connective tissue, the posture of rickets-injured children who also happen to have a congenitally weak connective tissue is particularly jeopardized. These children almost always suffer from foot problems and a curvature of the hip.

Talipes Valgus

Acquired, not congenital deformities

As with all poor postures and actual posture problems discussed in this book only those faulty foot positions will be included which have developed as a result of weak connective tissue, weak ligaments, and weak muscles — in other words, only those problems that are actually correctable through exercise. We are thus restricting our discussion to acquired deformations for the most part and not to congenital ones, which frequently require radical measures such as casts, braces, or surgery.

Talipes valgus (see the following illustration) is a very characteristic deformation of the ankle which often first catches the parents' attention when they notice that their child wears out the soles of his shoes unequally. The inner part is more quickly worn out than the outer edge. This unequal wear is in itself an

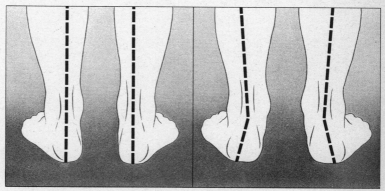

The normal ankle axis and the decussate position in talipes valgus. This illustration shows that the inward padding of the ankle joint recedes when the inner edge of the foot is raised.

indication of a wrong and disproportionate stress. The causes for this, as for almost all acquired faulty joint positions, are weak connective tissue, including the ligaments and the articular capsule, and missing muscular support. In this case, an inherited constitutional weakness and a deficient functional training usually coincide and result in the fact that the joint in question is very weakly held together. In the case of talipes valgus, the inner ankle joint protrudes heavily inward, and the joint axis is shifted. Parents can easily observe this shift by examining the line formed by the heel bone and the transition to the achilles tendon. In the case of talipes valgus, this line of foot and leg forms a clear inside angle which can actually be described as a decussate position of the ankle. If one lifts and at the same time supports the inner edge of the foot, the sagging of the child's ankles will disappear just as it does when the child sits and dangles his legs because now the whole foot is relieved of all weight.

This is already one indication that poor foot positions in a child are still "loose" and become apparent only under the stress caused by the weight of the body. When this weight is removed, they balance themselves out again. Since this is so, the poor position can be corrected by training the small foot muscles that together support the arch.

Faulty joint positions

Every faulty position of a joint affects the neighboring joints as a consequence of the body's efforts to maintain its overall statics. This begins below with the feet and continues upward until it reaches the spine. In the case of a poor inward position of the ankle, this deflection of the static equilibrium is also passed on upwards and downwards from joint to joint. This is why one rarely finds a faulty foot position alone; in the case of talipes valgus one usually finds a flat foot connected with it.

The main reason for the inward deflection of the ankle in the case of talipes valgus is a sagging of the longitudinal arch of the foot. Another, but rarer, instance occurs as a compensatory phenomenon connected with a child's bowlegs.

Flat Feet

Ligament and capsule weakness

This poor foot position, too, is the result of a foot weakness: in this case, the ligament, capsule and muscles are too weak to properly support the foot. The small foot muscles have the very important function of supporting and bearing the foot, for the foot is composed of a number of individual parts that are passively joined by ligaments and articular joints, but which are also supported by many small muscles. If their tensile strength abates, the total weight of the body rests upon the passive connective tissue portions of the foot, which slowly overextend themselves, give way, and let the arch of the foot collapse.

The architecture of the foot as a whole lends itself to comparison with an architectural arch whose longitudinal and transverse vaults absorb the building weight bearing down upon them. If they flatten out, not only the ankle sinks inward and the statics in the knee change, but the suspension of the body weight is also lacking while walking; the torso and the spine are subjected to more stress and they are jolted by a "harder" gait.

The essential stress points in the arch construction of the foot are the heels and the balls of the big and little toes. If one were to connect these points with lines, they would produce a triangle with the line running along the inner edge of the foot corresponding to the longitudinal arch, and the cross line running along below the balls of the toes forming the transverse arch, while the third side, the outer edge of the foot, has a fixed position. These arches receive their respective curvatures from individual component parts: the longitudinal arch from the individual tarsal bones, the cross arch from the metatarsal bones. They receive their shock absorbing elasticity and their cohesion from the ligaments, the articular joints, the tendons and the muscles. If the solidity of this framework should deteriorate, the arches flatten out, and the component parts of the joints which are arranged on top of each other slip almost onto one level. In the case of the transverse arch, this leads to a considerable widening of the top of the foot — to the flat foot, in other words, whose metatarsals now no longer lie on top of each other in arch formation but lie wide and next to each other. In the longitudinal arch not only the arch flattens out, but the next higher tarsis construction also begins to slip, which makes the ankle protrude inward. This then forms what is called talipes valgus.

Whether a foot is still largely capable of bearing weight, that is, if it can maintain its arch under stress, can be easily seen from a wet footprint left on dry stones. Almost everyone is familiar with these "pictures" at swimming pools and beaches. Almost no one, however, critically observes them in the case of their own children. Of course, one should also remember that an infant has a very fleshy foot, and this can simulate a flat foot without actually being one. Still, keeping this in mind, it is a good idea to check a child's footprints occasionally, for this is a reliable form of early detection of the problem.

Knock-Knees, Bowlegs

A child's knock-knees are one of the most frequent deviations in the area of the knee joint. Even though an infant has heavy, slightly bowed legs, when he stretches them full length one can already notice a knock-kneed position gradually coming into being.

The critical years

As a matter of fact, a very large percentage of children between the ages of two and five have knock-knees, and this is why specialists also speak of "physiological knock-knees" during this growth period. However, this condition should have largely regressed by the time the child starts school and should never exceed a certain intensity, for this deformation can also lead to osseous changes in the region of the growth sympheses, to stronger overextensions of the articular joint, and, above all, of the inner ligaments.

We have already pointed out that the pressure that weighs down upon a bone has an inhibiting influence on its growth. An especially vivid illustration of this point can be seen in an increasing knocked position of the knee joint. Once understood, one should take advantage of the functional stimuli of bone growth as well as of the bone's specific manner of reacting to pull and pressure in order to correct the growth direction of very severe knock-knees. This can be done through special treatment involving braces that fix the growth sympheses on the inner side of the knee and thus inhibit further growth in this region until the growth of the outer portion of the knee joint has had a chance to catch up.

Correcting the growth direction

The majority of cases of knock-knees in children, however, can be corrected through the more simple measures of redistributing the stress on the knee. This is accomplished by instep raisers or an elevation of the inner edge of the shoe. In many cases, this proves enough in giving the leg the correct position for growth.

Knee deformations that can be alleviated

A congenitally weak support apparatus of the articular joint and, more importantly, of the joint ligaments in the knee can be offset by means of a goal-directed training of the muscles responsible for the stabilization of the knee. We emphasize these so-called "inner reins" in the following descriptions of the exercises because of their importance for the support of the knee joint. Since the majority of knock-kneed children have weak connective tissue, they usually also suffer from flat feet or clubfeet and have difficulties in extending their backs.

The "back knee" (*genu recurvatum*), too, which actually bends backward, is almost always a consequence of a general weak connective tissue and weak ligaments. The treatment for this is similar to that for the knock-knee, with the important difference, however, of an emphasis on strengthening the thigh muscle (*quadriceps femoris*) which shares the responsibility for fixing the knee joint during extension.

We will not discuss various other possible knee deformations since they are not correctable through exercise or through other measures that parents might apply. These include poor joint positions due to broken bones, torn ligaments, tumors, or paralyses of one kind or another.

Bowlegs are much less common today than knock-knees. Their flexure occurs less frequently in the knee joint than in the osseous portion of the shin. The causes of bowlegs are often rickets or other bone growth disturbances, broken bones, tumors,or inflammations in the bone growth zones. Since they almost always have to be treated surgically, they will not be discussed in connection with our posture exercises.

Kyphosis and Lordosis

Deviations of
the spine

It is much more difficult when dealing with children than when dealing with adults to isolate and diagnose the various deviations of the spine. Many children thought to have posture problems may simply be demonstrating a variable posture: their stomach begins to protrude when they become tired, their shoulders become rounded if they have to stand for a long time, their shoulder blades protrude greatly when they cross their arms in front of their chest, and one can notice a curvature of the hip when they sit on the floor with their legs stretched out straight in front of them. All of these symptoms, however, disappear again when the child is rested: in fact, he can correct them himself when he is told to "stand up straight", for they are not yet fixed as in the case of most adults. The poor positions become fixed — and this also applies for joint deviations in other parts of the body — only when osseous changes make a correction of the faulty position impossible or, in other words, when a weakness of the support system is not the sole cause of them.

47

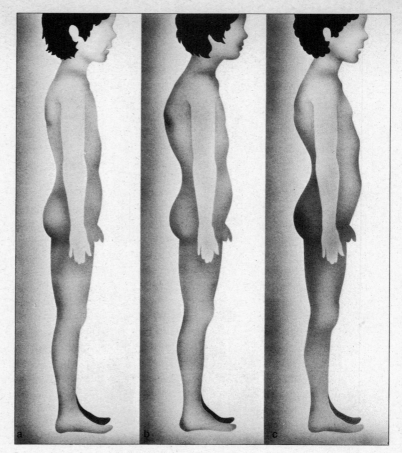

Comparison of normal posture (left) with the most frequent posture problems:
a) A good posture for a child with the normal physiological curvatures of the spine.
b) A hunchbacked child with rounded shoulders. A child can fall into this "tired" posture after standing for a long time.
c) Lumbar flexure with protruding stomach, which the child unconsciously uses to balance out a rounded back.

On the other hand, a child cannot stand perfectly straight for any protracted length of time because his connective tissue is too weak and his muscles tire too quickly. This is why there is no use in constantly telling these children to stand up straight. There is only one sure way to help them, and that is to transmit to them a feeling for posture and then to do the appropriate exercises with them.

All backward curvatures of the spine — those occurring in the

Instill a feeling for posture

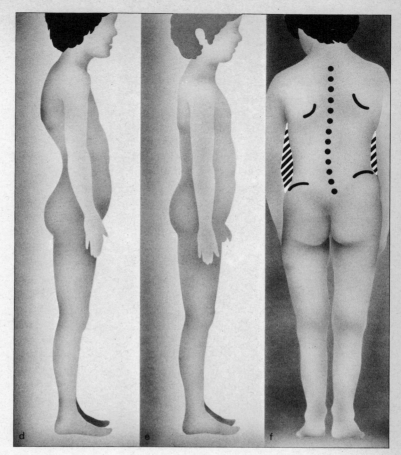

d) Round shoulders: rounded back, slightly protruding shoulder blades, lumbar flexure, protruding stomach.

e) In the case of a flat back, the normal physiological curvatures of the spine are almost completely missing.

f) A scoliosis is a lateral curvature of the spine. It can be seen primarily in an unequal height of the shoulder blades and of the iliac crest of the pelvis as well as in a lopsided waist.

convex portions of the spine — are called *kyphoses* (sometimes spelled cyphoses). Within certain limits they are normal constituents of the physiological curvatures of the spine and can be found over the sacrum and clearly marked in the region of the thorax. Forward curvatures of the spine, in other words, those occurring in the concave portions of the spine, are called *lordoses*. These are found in the lumbar region of the spine and in the neck. When these curvatures exceed the normal degree,

49

or are abnormally increased, one speaks of hunchback, swayback, or a curvature of the hip. The essential differences between all of these curvature formations are their degree of severity and their location in the various regions of the spine.

Curvature of the Hip (Hip Kyphosis)

This convex curvature of the spine with its vertex at the point of transition between the lower thoracic and the lumbar vertebrae can be seen very frequently in young children and is caused by their learning how to sit at too early an age. Often it can be increased as a consequence of an early childhood rickets.

Hip kyphoses rank among the most frequent posture problems in children. A great many adults also have a curvature of the hip, but few are aware of it because it is not visible while standing or sitting in a normal position. One way to determine its presence is to sit on the floor with the legs stretched out straight in front. In this position, you can see whether and just how far you can extend your back. If it does not quite manage to extend itself in the region of the lumbar and the lower thoracic vertebrae, it compensates by overextending that portion of the spine directly above it. This is the point where adults will feel backaches. They are the result of a hip kyphosis in childhood. Since it is so prevalent, this posture problem should not be minimized, especially since there are exercises that really can improve it.

Backaches in later life

The most important goal here is to strengthen the deep hip flexor muscles which run under the inguinal ligament toward the back and pull the pelvis and the spine forward from the stomach side. A precondition for the forward pulling of the pelvis is the extension of the iliopsoas muscles that run along the back side of the leg, because without this it is impossible to simultaneously stretch the legs and the back in an upright sitting position.

Weak hip flexors

As mentioned before, hip kyphoses can be caused by sitting at too early an age (recall what we said before about children's walkers), and are ususally connected with rickets. They can be caused as well by a generally weak connective tissue system combined with inadequately developed back muscles.

50

Hunchback (Round Shoulders)

The previous section has pointed out that, unlike adults, children do not yet have real hunchback. In the case of an adult, the formation of osseous, wedge-shaped vertebral deformations eventually preclude the complete extension of the spine. A child's hunchback (or round shoulders), on the other hand, appears for the most part as "tired posture" after the child has been standing for a long time. Fatigued, the child unconsciously tries to compensate his drooping round shoulders by means of a corresponding lumbar flexure. When dealing with children, then, the whole spectrum of transitional forms of this posture problem ranging from hunchback to round shoulders can be detected, and it is very difficult to isolate one form from another. Children with this type of poor posture stand with forward falling shoulders, slightly protruding shoulder blades, a round back, a lumbar flexure and a protruding stomach. They can quickly correct this posture when told to do so by straightening up, but they immediately sink back into it if their attention is diverted.

"Leaning" over school work or in front of the television also fosters this kind of round-shouldered posture. If nothing is done to correct it and it becomes a lasting habit, the front chest muscles, especially those that form the edge of the armpits, become shortened and pull the shoulders even further forward.

The most important goal as far as exercises for this kind of weak back is concerned is thus training the muscles that extend the back as well as stretching the front chest muscles. This is the only way an extension of the back and a pulling back of the shoulders can be made possible at all.

Another essential consideration in this overall context is the proper dimensions of the child's working area. The table or desk at which the child sits to do his homework, for instance, ought to be proportionate to his size in order to encourage him to sit up straight. Lying on the stomach in front of the television should replace sitting; but the most satisfactory solution to this posture problem still remains a strict restriction on the time spent watching television in the first place. Almost all of these examples of poor posture among children are due to a deficient support tissue, weak ligaments, and inadequately developed back muscles. These problems, however, are considerably

Proper working area

51

intensified by rickets, and many of our children with possible posture problems can be considered as suffering from slight borderline cases of this disease.

Lumbar Flexure

A lumbar flexure is hardly found by itself as a posture problem. Like all of the examples described so far, it, too, is a habit-formed poor posture, but, in addition, it also unconsciously compensates for a rounded back. When a child with a usually straight and strongly extended back slumps because of fatigue, all the normal curvatures of his spine are necessarily intensified: the back becomes rounded in the thoracic region, draws itself in in the lumbar region and forms what is called a lumbar flexure, and actually becomes shorter. As a matter of fact, children with poor posture due to fatigue are usually about an inch (several centimeters) shorter than when they are told to stand up straight.

Compensated rounded back

The most important starting point to correct a lumbar flexure is to lift or raise up the pelvic girdle. This can be done, however, only after the child has acquired a feeling that he has a pelvic girdle and that he can consciously move it. One way to transmit this feeling to the child is to have him stand with his back against a door frame or against a wall and press the small of his back flat against it. This is a frequent part of the exercises described later and should be practiced whenever the child tends to slump, whether during the exercise period or at any other time.

Awareness of the pelvic girdle

Protruding Shoulder Blades

Protruding shoulder blades are sometimes called angel wings (*scapulae alatae*), a very apt description of their appearance. In some cases they protrude to such an extent that one can actually grasp beneath them. This clearly shows how flexible and weak the tissue in this region is.

Angel wings

The main cause for this poor posture picture is a weakness of the subscapular muscle and of the muscles between the shoulder blade and the spine. This can be demonstrated by watching a profile view of the child's back while he does a simple exercise. The child should sit down, on a chair or on the floor, and put his hands on his hips. Once in this position, he should press his elbows back against his mother's hands in such a way that a tension but no movement results. This makes the muscles between the spine and the shoulder blade tighten and the protruding shoulder blades disappear.

Strengthening wings

Since these muscles are very weak in children with "angel wings", they allow a continuous sideward slip of the shoulder blades which, as part of the shoulder girdle, give way to the pull of the arms that are hanging from them. This displacement leads to a constant overextension and weakening of the shoulder blade muscles and to a rounded forward-hanging position of the shoulders.

Protruding shoulder blades are a clear indication of the connection between muscle strength and their ability to pull the shoulders back — in this case the muscles are too weak, thus allowing a passive sideward slide of the shoulder blades. In extreme cases, these may even stand out in relief against the ribs.

Strengthening the shoulder blade muscles

The primary therapeutic goal in this instance is the strengthening of these shoulder blade muscles. One way to do this is to support oneself on one's hands and arms. All push-up and propping exercises are thus appropriate here.

The Flat Back

In light of what has so far been said about kyphoses and lordoses, the layman might be tempted to be completely satisfied with a flat, perfectly straight back without exaggerated curvatures. As a matter of fact, many people do not consider a flat back "poor posture". If we take a closer look at children with this posture problem we can see that because of the steep and erect position of the pelvic girdle, the physiological curvatures of the spine are almost completely equalized, and the vertebrae are therefore positioned one above the other like a stick of bamboo. Another conspicuous aspect is that the back muscles, particularly those between the shoulder blades, are poorly developed.

The spine lacks elasticity

This has two functional effects: first, this kind of stick-like spine with the vertebrae one above the other loses its shock-absorbing elasticity. It is more or less severely jolted with every step. During the course of a lifetime, therefore, it is quickly worn out. Secondly, the weakly developed back muscles, which are constantly "spared" by this flat posture, are much less fit and tire very easily. Many back problems in adults are the consequence of this kind of flat back, whose discs are prematurely worn out and whose weak muscles cramp because of overexertion and eventually cause pain.

Strengthen the back muscles

It is difficult to prescribe a treatment for a flat back because one cannot simply train a child to develop a lumbar flexure and then hope that the spine will bend to correspond to it. The only thing that can be done is to try to keep these children's backs as elastic as possible and to strengthen their back muscles, especially in the region of the upper thoracic vertebrae. Thus, for this particular posture problem all those exercises are appropriate which contribute to the strengthening and the agility of the torso, particularly all the bracing and propping exercises designed for the shoulder blade musculature.

Scolioses (Lateral Curvatures of the Spine in Various Regions)

Scolioses are lateral curvatures of the spine, but since they also show the normal forward and backward flexures of the back they usually appear as curvatures in several regions at the same time. Scolioses can be found in all regions of the spine. They may involve only one large curve (called C-Scoliosis) or they may exhibit several small curves, which then show up in an S-formation.

C-Scolioses, S-Scolioses

S-shaped scolioses consist of a "primary flexure", the main

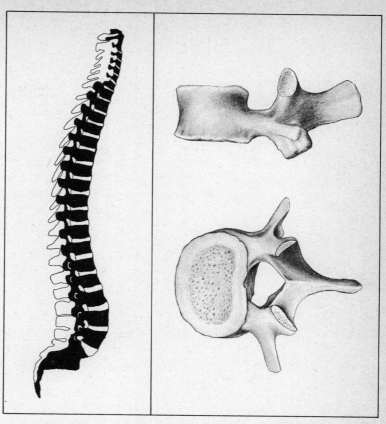

The spine with its normal physiological curvatures. The individual vertebrae (*left*) are positioned on top of each other in a roof tile fashion. The boney protrusions form the "bumps on the backbone." Individual vertebral bodies (*right*) as seen from the side and from above. The acantha is the spiney projection on the right of each illustration; in the lower illustration the boney spinal canal is shown between the acantha and the vertebral body. The spinal cord runs through this canal.

curvature, while the other curvatures ("secondary flexures") are an attempt on the part of the body to maintain its statics and symmetry. The primary flexure is fixed at a relatively early age and thus cannot be corrected, while the subsequent equilibrium-maintaining flexures remain flexible for a longer period of time.

One essential characteristic of the lateral curvature of a scoliosis is always a twist, and this is what frequently makes any treatment of the condition very difficult. On the other hand, it is this twist that draws attention to a slight scoliosis in the first place, for the individual vertebrae, which are also twisted to the side, make the ribs of this side of the chest buckle out somewhat, while the other side of the ribs is drawn slightly inward. If a child bends forward in a relaxed position, one can usually recognize the presence of a slight scoliosis on the asymmetry of the surface of the back.

Scolioses should always be examined by a specialist, for it is only through an x-ray that one can determine the true degree of the curvature and the twist of the spine. The latter can frequently be much more severe than the slight deviation of the line of the backbone might suggest. They must also be reexamined regularly, since they often intensify greatly during the various growth spurts in a child's development. This is particularly true for puberty.

Scolioses can be congenital, but they are more frequently acquired. We will limit our discussion to those acquired scolioses that might possibly be prevented through attentive observation and appropriate correction. Falling into this group are all those scolioses which have developed as a result of a disturbance of the body symmetry. These disturbances include a crooked position of the pelvis, a shortened leg due to a paralysis resulting from polio or an injury, unequal knock-knees, a congenital dislocation of the hip, or injuries to the vertebrae caused by rickets. The one-sided stress brought about by carrying school bags on one shoulder presumably plays a lesser role in the formation of scolioses than one used to assume. A child usually unconsciously changes hands when carrying something simply because his hand or arm gets tired. However, this is a very different problem in the case of children previously injured by rickets as well as in the case of those children with weak connective tissue and an inadequate back musculature. Here a heavy school bag that is primarily carried on one side surely can lead to a scoliosis; at the very least it can intensify an already existing slight scoliosis.

What might seem as undue emphasis on scolioses has its purpose: many parents frequently make too light of this condition, mainly because they cannot accurately appreciate the degree of the curvature and they thus become very alarmed when with the beginning of growth spurts during puberty a severe curvature of the spine and ribs appears "all of a sudden." This is precisely why scolioses have to be treated as early as possible and with a very consistent program of physical therapy aimed at training and strengthening the muscles that extend the back. If this is done properly, they will then be able to maintain the total extension of the back and thus effectively counteract the tendency toward an intensification of the curvature.

Since this posture injury is a serious one, and in order not to lead parents into the temptation of devising an exercise program on their own for a child who has an apparently harmless scoliosis, no appropriate exercises have been presented in this book. This posture condition ought to have professional orthopedic treatment and physical therapy.

Funnel Chest, Chicken Breast

Funnel chest is the name of a more or less deep depression of the breast bone (sternum) toward the spine. It occurs very frequently in connection with other deformations and very often with scolioses. Since it is congenital, medical science usually ranks it among the group of "congenital deformities", although fortunately most of the children afflicted with it do not feel actually "deformed". However, it is not, as is frequently assumed, a consequence of rickets. It should be treated by a specialist as early as possible, and the prescribed exercises should be continued at home as well, since this condition can be improved by consistent treatment.

Early treatment is necessary

The treatment for funnel chest aims at raising the breast bone (sternum) through an expansion of the chest which is induced by breathing exercises. Before starting this type of exercise program, however, the adult should be carefully instructed in how to direct breathing exercises, for a lot of harm can be done through forced breathing. The exercise section of this book also provides suitable expansion positions, but, like most of these exercises, they are only of value when practiced conscientiously on a daily basis.

The condition called "chicken breast" exhibits a wedge-shaped breast bone that is arched at the top. Its origin is rickets and it is not correctable through exercises. Nevertheless, it can be gradually improved through pressure bandages as long as the bone is still pliable. This, however, is the sole responsibility of a physician!

Congenital Dislocations of the Hip Joint (Hip Luxation)

The clinical picture of a poor formation of the hip joint has no place in our exercise program simply because it cannot be treated by laypeople. Still, a brief discussion seems justified because it is such a frequent problem. Perhaps the most one can do is to sharpen parental attention and awareness of this hip joint disease. The disposition to a hip luxation, an abnormal dislocation of the hip joint, is inherited. It does not turn up in every generation, but it does appear repeatedly at irregular intervals. Families with a history of hip luxations should therefore be particularly attentive to indications of this abnormality in newborn children.

Girls more frequently afflicted

Girls are four to six times more frequently afflicted by hip luxations than boys. The only way to accurately determine the existance of a hip luxation is an x-ray taken when the ossification centers at the head of the hip joint are still forming, in other words, at the age of three to four months. It goes without saying that one should not unnecessarily expose an infant to x-rays, especially if there is no strong suspicion of a hip luxation.

The first stage of this kind of change in the hip joint consists of a poor formation of the acetabulum (the socket-like depression which receives the femur; it is also called the cotyloid cavity) whose roof is too flat. The stress of standing and walking is what first brings the actual dislocation into existence, the slipping of the head of the joint of the thighbone out of the cotyloid cavity. This dislocation, therefore, first becomes visible only after the first year of life.

Growth stimulus via braces

The treatment of this disease consists primarily in stimulating the cavity toward growth around the head of the joint, to create the missing growth stimulus, in other words, and this happens through the aid of splints and braces. Horizontally spread out thighs exert the lacking stimulus on the cavity which can then normalize and round itself out to a great degree. The fact that this is actually a suitable means of regulation has been shown by studies done among tribes of peoples who carry their children on their hips. The child's legs are thus spread out horizontally, and a hip luxation is almost unknown. Of course, the most important thing is to recognize the poor position of the hip and leg in plenty of time. It can be seen on the unequal folds on the knee or the bottom, in a certain lack of mobility, or in some trouble in spreading the legs. Only a physician should perform these examinations, since every attempt at overcoming a mobility obstacle in an infant can itself lead to severe growth disturbances of the bone. Severe cases of hip luxations require various plaster casts, and a discussion of these procedures is at best peripheral to our topic.

EXERCISES

6. Necessary Preliminary Information

Why exercise at home?

It is important to note that the explanation of the various changes in a child's posture, both in the region of the feet and legs as well as in the region of the back, are by no means intended to encourage the layman to treat already existent real posture conditions himself based upon a presumed knowledge of the field. Every posture condition has to be orthopedically evaluated since a layman cannot assess the degree of the harm done; moreover, he must, at least in the beginning, become familiar with physical-therapeutic aspects of the exercises that are specifically designed for him. Only then can and should the layman undertake an exercise program at home. The purpose of the exercises presented in this section is to bridge the gap between visits to the specialist.

Determine
the degree
of injury

The exercises described in this book are not intended to turn mothers into perfect physical therapists, nor can they heal pathological changes in the spine! However, they are supposed to prevent any real injuries that can result from posture problems and they are supposed to strengthen and help those children who are constitutionally prone to posture problems to train meaningfully. Finally, they may also serve as a reminder for physical exercise at home in connection with an individualized, goal-oriented treatment program.

Since the course of movements is often difficult for a layperson to imagine or reconstruct in his mind, the suggested exercises that follow necessarily have to be limited to relatively simple ones. The wrong execution of an exercise is not only more strenuous than need or should be, but, more importantly, it can also do damage. Finally, it must be pointed out that there are always physicians who reject goal-directed orthopedic posture exercises because they do not always satisfy the lively motor drive of a child. This is undoubtedly true, as long as the child is fortunate enough to be growing up under ideal conditions, namely, with enough time and room. Minimum prerequisites for this are a large apartment or house and a large yard with gymnastic equipment available. By large we mean that the child should not get caught in the neighbor's washline while swinging, always supposing he has time to swing in the first place. Of course, ideal conditions such as these do not exist for most children, and as long as this is the case, there will be a constant need for goal-oriented exercise programs aimed at improving a child's posture.

Individual or group exercises?

Whether the exercises should be done individually or in a group depends to a great extent on the physical development of the child. It is very important for all concerned to understand this before starting an exercise program, for a lot of harm can be done through a wrong evaluation of the child's needs.

Exercising with infants and preschoolers:
An infant does not need a group; as a matter of fact, it prefers to play by itself. It does not need competition with peers, but instead praise and recognition from its parents. This is why a parent or a close person with whom the child gladly identifies should ideally always participate and demonstrate the exercises to the child, for at this age the child still learns through imitation.

Thanks to their lively imagination, all children completely identify with their respective role; accordingly, some exercises call for a dog that "barks", a bear that "growls", and a pony that "whinnies" while wildy shaking his legs. Since all children also

64

enjoy praise, recognition, and a bit of admiration, they can attain these things by performing various "tricks".

One good way to ensure this is not to exercise "abstractly" with little children, but rather to always start out with the attitude "who can…" followed by great and marvelling recognition of the success. The tricks, of course, also have to be within the child's reach, for failures are difficult for children of this age to bear. Curiosity is also strongly marked in very small children and can be profitably transformed into movement and gymnastics.

An important element in exercising with very young children is the inclusion of interesting objects or toys, a large and colorful ball, for example, or a training bag. Frequently a change in the room, such as cushions and pillows lying on the floor or upturned chairs and tables, also contribute novelty and enhance the interest.

Exercising with school children:

Exercising in a group

By the time the child is old enough to start school, he is at an age in which a group or a community has a fostering effect: it spurs him on when several children his own age are exercising together. Interest in competition also begins at this age, but it is not yet so marked that the individual child will try to outdo the other. Nevertheless, the adult in charge must see to it that they do not bicker about the role of leader, for children can very frequently become inconsiderate of each other when it comes to who shall be first. Even so, they usually demonstate a willingness to follow the suggestions and directions of parents and are pleased when other chldren also respect their parents. And above all, children strain to gain their parents' love and praise, and only afterward to acquire recognition in their peer group. When a group of children is exercising together it is extremely important to repeatedly draw attention to the accomplishments of each child.

For the child to learn and realistically assess his own abilities in comparison with those of others, there should be a list of common fitness standards against which he can measure himself. This also fosters an ability to criticize oneself. If the opportunity of a real competition among peers should then present itself, a situation in which being accepted in the group is more important than the recognition of one's own parents. Those children who do not have a realistic appreciation of their own abilities become increasingly less competent as "gymnastic leaders". This is the time when the orthopedic gymnastics should be either shifted into a therapeutically directed gymnastic group or else the common gymnastics at home should be continued with a new emphasis to meet individual needs.

The parents as partners

This new emphasis includes a clear role change on the part of the parents from demonstrator to the participant. If the child, even an older child, senses that his parents are doing the exercises for their own good as well, that it is equally necessary for all of them, whether in order to get rid of a "tummy" or to stretch out desk-induced round shoulders, he is drawn into the common effort. Moreover, it will frequently turn out that many exercises are far easier for the child to perform thanks to his greater agility than for his out-of-shape parents. The child should be allowed to sense and enjoy his superiority in this. It compensates for many of the personal defeats that the child has had to accept from the adult world.

The right time

It is most important that gymnastics be fun for children. They should not be perceived as a burdensome chore. Exercises should always have an aspect of "reward," a pleasurable daily event in which everyone "is allowed" to participate. This is why children should have all their other daily obligations and duties already behind them before they start.

The exercise time should also coincide with the time of day when the child experiences a particularly strong need for mobility. Children, too, have a daily rhythm with times of particular activity. By taking advantage of this, parents can employ the child's roughhousing urge, which seems senseless and often annoying to them, and channel it meaningfully.

Meaningful channelling of motor drive

If the child is nervous and fidgety, the program should start with active exercises that consume excess energy, so that afterwards the child will be better able to concentrate on the goal-oriented exercises.

One high point of a child's motor drive occurs in later

afternoon, usually just before the evening meal. This is also a favorable time from a health standpoint because the stomach is still empty and the child can move about easily and unhindered. Equally important, however, is the fact that the daily chores are out of the way.

Doing gymnastic exercises at this time also has the great advantage of stimulating the appetite — supper or dinner tastes especially good after a work-out. Another advantage to this time is the fact that the child will tire naturally and will thus sleep better.

The right amount

The time spent on exercises should not be too long. If it is, there will be difficulty in building enthusiasm for the next time. The child should also be left with some sense of regret that the exercises are already over for the day.

A quarter of an hour usually suffices for the gymnastic period.

Stop while it is still fun

If the children are not tired and are participating actively, they can continue for up to half an hour. The length of time, however, should always depend upon the child's mood at the time and his need for mobility. It is better to stop too early once in a while than too late, when the child has already begun to be bored.

The best plan is to set aside two evenings a week as gymnastic periods. A certain regularity pays off in the end. But it is better to skip a day after a few weeks and then begin again with a fresh start and renewed enthusiasm than to debate each time whether or not one feels like exercising.

The kind of exercise

Plan your program

The organization of the gymnastic period should be planned in advance by the mother or father. Do not underestimate the value of your imaginations here. If lengthy interruptions occur, the child slips from your control and his attention and enthusiasm wane. It is always a good idea, therefore, to choose and practice a few suitable exercises before you start so that the actual exercise period can proceed smoothly. Only then will the child take his gymnastic exercises seriously.

This exercise period, just like a meal, should have an accentuated beginning and an equally accentuated end and should not "fade" away. A good way for all involved to begin is to arrange the room and change their clothes.

The exercise that ends the period should be a little special and should stand out from the rest. One way to do this is to once again thoroughly stretch in an upright position, raise the arms, stetch up to the fingertips and slowly swing the arms in a semicircle back to the original position. A distinct end creates the feeling that something has been accomplished and finished.

Plan your program in such a way that the starting positions of the exercises do not have to be changed too frequently. Start, for example, in the supine or stomach position, and after a few exercises proceed to the upright sitting position with stretch legs or the crossed-legged sitting position, then move on to the all-fours position and finally stand up erect. This progression lends a better transition between exercises and each subsequent one develops organically out of the previous one.

Make critical comments about the exercises: praise or correct, but do not simply accept the child's efforts without comment. Children need to feel that they are constantly the focus of their parent's attention, and nothing dampens their enthusiasm more thoroughly than if their efforts go unnoticed. If several children are present, each one should be repeatedly addressed by name and recognition and correction should be fairly distributed. Of course, no one should ever laugh at a a clumsy child, but this goes without saying.

Room and clothing

Choose a room that is large enough for an adult to roll backwards without bumping into anything or tipping anything over. It should also not contain any fragile objects that have to be constantly watched and protected. Remember to include furniture and other objects, cushions for example, in the exercises whenever possible. If there is no rug in the room, a woolen blanket can be folded a few times and still remain large enough for a person to stretch to full length on it. It is not necessary to have a pillow under the head.

Barefoot
is best

Foot exercises are always done barefoot. In fact, barefoot is the healthiest way to do any gymnastic exercise. Children catch colds less easily than most parents believe. If they really are susceptible to colds, the best means to ward them off is to strengthen the child's resistance by getting him used to exercise in the open fresh air. During the exercise periods, for instance, the windows should be open and the participants should not dress too warmly. Pure cotton undershirts and underpants are the most suitable. Children get warm more quickly than adults do. If the room is cool, the exercises should start off with vigorous hopping around as a warm-up exercise. If there is a garden or a backyard available, the exercises should be done outside as often as possible.

Appropriate equipment

Gymnastic equipment can be used in many ways for many different purposes. Furthermore, they are cheaper than many unimaginative technical toys now on the market. Often large sums of money are spent on sophisticated electric toys, while the child would have profited much more from and have done many more things with a large inflatable training bag.

Purchases
that pay off

Moreover, appropriate equipment lends much more weight to the exercise period. An athletic wand looks a lot sportier than a broomstick, which could also be used for many exercises, but if the broomstick is used the child gets the feeling that the whole program is improvised. It is a good idea to purchase several such wands so that several children as well as the mother can exercise at the same time. These wands are not expensive. It also pays to buy several jump ropes and gymnastic balls, which are easier to grasp than normal balls and stay inflated longer.

Many other pieces of equipment were first developed to help handicapped children by stimulating their motor function. Their use, however, is obviously not limited to handicapped children, for healthy children can also profit from exercising their imagination in using them in new and unique ways. Some of these items are a large, inflatable training bag upon which one can hop, ride, and lie on one's stomach, a large inflatable but stable ball, a wooden swing and a saddle bench, which can be taken apart and used for other things such as steps. There are also top-like objects that provide various opportunities for balancing exercises, and a trampoline is always a source of especially great fun. All of these things, however, should be used *only* under parental supervision. They are especially good for strengthening body tension and agility as well as for training and meaningfully directing a child's motor function.

Since most of these pieces of equipment can also serve as furniture in a child's room, there is no reason why the buying of them should involve any extra expenses. On the contrary! A child's room can be arranged with a view towards function, so that it meets the child's mobility needs. Other equally suitable and inexpensive objects for a child's or exercise room are foam rubber cushions of different sizes which can be covered in colorful and durable materials.

Basement as recreation room

I recently saw a basement room which was sensibly set up not as a bar, but as a recreation room for the three boys of the family.

Among other things it contained a hammock that was fixed to the ceiling, a rope ladder and a thick climbing rope, both of which were also hanging from the ceiling. On the floor lay thick and colorfully covered foam rubber cushions which formed part of the brightly painted houses the children had built out of large cardboard boxes, the kind used for packing washing machines and similar objects. They were large enough for the children to crawl into, around, through, and out again. This detailed description is not meant as a prescription, but rather as a suggestion in the hopes that it might provide some ideas for the future arrangement of a child's room. So little thought usually goes into this, and yet it is extremely important that the child's room meet and stimulate his creativity and his need for mobility.

Suitable sports

When thinking about the possibility of an appropriate program of physical training, the question naturally turns on the one hand to those sports that can be played alone or together with the family, as well as to those sports that can be played by groups of children of the same age. The important thing to be considered here is whether the sport is particularly aimed at improving the child's *posture* or whether it is primarily concerned with stimulating *movement* and *circulation.*

Swimming stretches the back

If an improvement of posture is the goal, swimming is the best possible sport. It's greatest advantage is that it fosters an extension of the back. In addition, the crawl stroke strengthens the muscles of the shoulder girdle. Rowing is also a good exercise regimen for the back, especially when it is done not only with the arms, but with the whole torso. Another possibility is gymnastic groups, such as those offered under the auspices of adult education programs. Many of these programs also concentrate on children's posture problems. Courses in modern ballet schools also give valuable posture instruction at the beginning levels. Basketball and volleyball can help improve posture, too, since these games frequently require the players to jump in a fully extended position. Similar ball games using a rope stretched across the yard can easily be improvised and children respond to them very enthusiastically. Bicycle riding and football or soccer, on the other hand, are primarily aimed at improving circulation and endurance fitness, and they are thus less effective for posture. Rollerskating, skiing and ice skating likewise foster a child's agility, endurance, and dexterity.

A new and interesting competitor in the field of posture and personality improvement are the asiatic sports that are becoming increasingly popular in our culture. These include, among others, Tai chi chuan and karate. Many cities have already set up courses for children and adolescents in these regimens. This fitness training program has as its goal not only enabling the person to defend himself, but more importantly, uniting a very systematic posture and tension training with breathing and concentration exercises. This form of concentrated bodily control surely does a lot of good for many of our unconcentrated, nervous children with posture problems. Another advantage is that it also strengthens self-confidence via posture training. It is advisable, however, to be thoroughly informed about the quality of these courses or of the course instructor before starting the program, for there can be considerable variations here.

The Starting Positions

To avoid repeating the explanation of the starting position for each exercise, this section will describe all the starting positions needed for the suggested exercises discussed in detail. Later it is important to carry out even apparently secondary instructions within the description of an exercise as precisely as possible, because the indicated rotation of a hand or a given position of the arms with reference to a definite part of the body, for instance, will produce a reaction that is essential for the exercise and its success.

Parents
should check
posture

Poor posture habits, such as overstretched joints, protruding shoulder blades and a lumbar flexure, can be seen and assessed very well from the upright standing position. Watch your child's posture and movements especially at those times when he is unaware that he is being observed, since this is when the weak spots are especially easy to notice. Also, if you suspect your child might have an unequal posture, check to see if the pelvic girdle is straight. This can be done by lightly pressing the inner edges of your hands on the right and left sides of the child's pelvis, and then right and left along the crest of the pelvis. Compare whether your hands are at the same height; also compare the length of the legs while the child is standing up. If you find any asymmetry here, it may be that the statics of the back which rest upon these regions is also not completely symmetrical. It is also possible that the spine may show a curvature, which is usually associated with a twist (scoliosis).

The shape of the spine can be seen most clearly when the child slowly bends forward in a relaxed manner. Stand behind him or her and look flat over the child's back. By bending the back, the row of vertebrae become increasingly clear, and deviations toward the side will become visable as will also an unequal, one-sided curvature of the chest cavity in the region of the spine associated with a twist, indicates that the spine is pressing the ribs outward on one side while the other side is correspondingly flatter.

One should not attempt to cure this kind of spinal deviation oneself, for special tension and posture exercises are necessary which correspond to the direction of the respective curvature.

Consult your
orthopedist

Every child with this posture problem ought to receive physical therapeutic treatment under the supervision of a physician. Puberty and its early stages is another important time to carefully observe a child's posture, for scolioses frequently remain "silent" for a long time and then worsen suddenly and considerably at the onset of puberty.

The following suggestions for exercises represent only one possible approach for improving a child's strength and posture. They are meant to encourage children in their joy of mobility and either to prevent posture problems or be an ongoing therapeutic treatment between visits to a specialist. If the posture problem is a serious one the gymnastic exercises should be constantly supervised by a specialist in the field.

Supine position:
If nothing is stated to the contrary, the legs are parallel, the arms lie loosely next to the body, the head is flat on the floor. If your head is uncomfortable, a folded towel can be placed underneath it.

Stomach position:
The legs are stretched out straight and lie parallel to each other, the heels fall slightly apart, the arms lie next to the body. Since the shoulders fall forward in this position, the arms turn somewhat inward by themselves, so that the hands lie flat with the palms upward. The head is to the side. In exercises aimed at stretching the back, the forehead should rest against the floor.

Side position:
The body is on its side, the stretched legs rest on top of each other, and the torso is well stretched. The head rests upon the extended lower arm, while the other arm is used to support the body by resting the palm of the hand on the floor in front of the chest.

Upright sitting position:
Sit on the floor with both legs stretched straight out. The torso is stretched so that the hip joints form a right angle with the floor. This can only be done if the pelvis is pulled far enough forward.

Spread leg sitting position:
The same position as in the upright sitting position, but this time spread the outstretched legs apart. This makes the lifting and forward pulling of the pelvis much easier.

Cross-legged sitting position:
The legs are bent as far as possible at the knee and crossed over one another, the pelvis is pulled forward, torso and head are extended. This position is easier the flatter the legs lie on the floor. And again, this depends upon the extendability of the adductor muscles.

Heel sitting position:
Kneel on the floor and rest the weight of the body on the heels, whereby the tops of the feet rest flat against the floor and the torso and the head remain extended. This tension causes some discomfort after a short time for those not used to it; if this is the case, place a folded towel under the knees and feet. Be careful not to fall into a lumbar flexure.

Kneeling position:
Kneel down with extended upper torso, extend the hips. The lower legs make a 90° angle with the rest of the body, the feet once again rest on the floor with the soles turned upward.

All-fours position:
Lean on hands and knees so that the hips and the knee joints form right angles. The feet rest on the floor with the soles upward. The back is as straight as possible — the small of the back may neither curve upward, nor may it sink toward the floor. There should also be no indentation visible between the shoulder blades; the head and neck should be held at the same height as the back. The arms are stretched straight and locked at the elbows, the hands are turned slightly inward in order to avoid an overextension of the elbows.

Upright standing position:
The feet are parallel and as close as possible without letting the ankle bones touch one another. The knees are stretched, but not overextended (this direction is frequently misinterpreted). The pelvis should not tip forward, for this would form a lumbar flexure and the stomach would protrude too much. Furthermore, the small of the back would also not be able to bear and balance the torso freely. The shoulders are held back with the shoulder blades flat against the torso, and the head should be held high without pushing the chin forward.

7. Exercises for Preschoolers and School Children

Romping Exercises for Warming up and Working off Excess Energy

The following exercises and games are primarily intended to meet the child's great motor drive. They can easily be scheduled at the beginning of an exercise program and be repeated now and again whenever needed. They can also end the program, especially when the child exercises in the evening before going to bed, because they then contribute to a sleepiness. In any case they should help encourage the child to work out correctly at least once a day, to stimulate his breathing and circulation, and to include all his muscles in the activity. The result is a healthy fatigue, free of all nervousness and fidgetiness. This helps the program especially in the case of school children who have to sit still all day. If at all possible, the exercising should be done outside during the summer. If not possible, the whole apartment should be utilized including up- and downstairs; the greater the outlet, the greater the fun and success.

Running barefoot through the apartment is always great fun. Let the child set the pace by being the leader, then the mother, and the child again. The chase goes around and under the table, over an easy chair, from which the smaller child can perhaps already hop down by himself, and, if available, up and down the stairs as well. It is better if the mother follows the child upstairs rather than downstairs. Remember, too, that what is demonstrated has to be imitated: for example, hopping with both feet held together as if they were bound, then hopping a bit on one leg, which of course can only be done properly if the child is at least four years old; running on tiptoes with arms raised high so that the child can grow as tall as a giant and then immediately thereafter crunching up like a drawf who has a great deal of difficulty moving forward. This slowdown is immediately compensated by giant steps as if the child were wearing seven-league-boots. The goal is to really get out of breath. After having thus burned off the excess energy, all participants return to the room in which the gymnastic exercises will be done.

In this exercise the child pretends to be a frog that is hopping away from the stalking stork behind him. The child should jump out of a crouching position and widely stretch both arms and legs in great leaps while mother or father follows behind with high, stalking steps.

This exercise is called the jumping jack: jump high and spread your legs sideways while clapping your hands high over your heads, and then vice versa: let the arms fall to your sides while the legs close together.

In this exercise the child pretends to be falling in the water, but there are pillow islands all around and he can rescue himself by taking great leaps from one island to another. The pillows have to be arranged far enough apart so that the child has to strain a bit to reach them without falling in the "water". On the other hand, of course, they have to be within reach. Another possibility is to build a footbridge, but the child must be careful to keep his balance from one step to another. This bridge can consist of a row of books, or, more difficult, a broomstick.

Stubborn ponies stand on all fours and kick their back legs out wildly. The object of this exercise is to see if Father can catch one of the "pony's" legs, hold on to it, and perhaps even lift the "pony" off the ground by these uncooperative legs.

This exercise resembles the starting position of a foot race runner. The child starts out on all fours and thrusts first one and then the other leg backward in such a way that one leg bends while the other one stretches.

There seems to be another frog in the house. All the participants jump up out of a deep crouch into a full stretch, let their arms fly high over their heads and then fall back into the crouch. This should be done a few times without stopping.

Open road! — All participants take advantage of the fact and lie stretched full length on the floor with their arms above their heads. Keeping stiff and straight and maintaining the tension, see who can be the first to roll back and forth across the room.

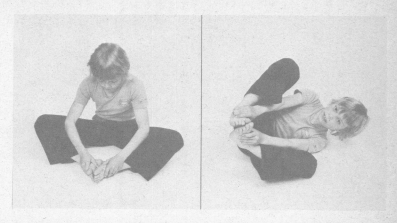

For this exercise everyone sits almost in the cross legged postiion, but does not cross his legs. Instead, he presses the soles of his feet together and holds them together with his hands. The next step is to swing yourselves toward one side, roll like a ball on your back to the other side, and thus return to the original sitting position. If you take a good enough swing, you can really roll like a ball all around the room or all over the grass.

It's time to jump again. This time place at least two jump ropes or two broomsticks side by side on the floor or grass and jump over them sideways from one end of the length to the other and then back again.

Several chairs or benches can serve as a "race course" by arranging them at certain distances so that the child can climb up on one, then hop down and climb up on the next, or climb over one and crawl under the next. They can also run around the chairs in large curves, hop around them on one leg, or crawl around them on all fours.

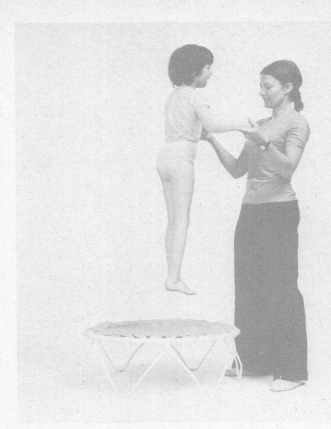

Jumping on a trampoline is great fun at any age and is an excellent way to train. However, it should be done only under supervision; if the apparatus is not being used, it ought to be put away.

Since both the romping exercises as well as the foot and knock-knee exercises differ very little for preschoolers and school children, all of these exercises have been grouped together. However, the torso exercises that place higher demands on bigger children have been treated separately in chapter 9, although young siblings can keep up amazingly well simply by imitating what they see the others do.

Exercises for Talipes Valgus and Flat Feet

Goal: *to strengthen the small foot muscles which stabilize and support the longitudinal and transverse arches of the foot; to tense the arch and increase the mobility and agility of the foot.*

Grasping exercises with the toes:

Use your toes to pick up small pebbles, buttons or marbles from the floor and place them in a box or bowl.

Pick up a pencil or a small stick with your toes, take it with your other foot or pass it on to a partner. Mother and child can also try to take a longer stick or a cooking spoon, which they hold in the toes of both feet, away from each other. Of course, you can only use your toes to do this.

Use your toes to pick up a handkerchief or a wash cloth and rub the other leg with it from bottom to top.

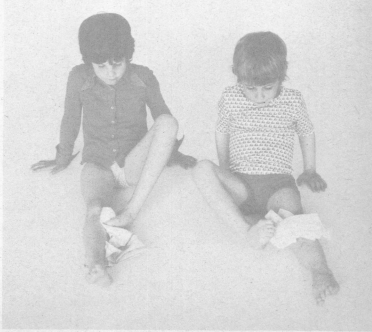

Use your toes to wrinkle up an unfolded towel and then carefully smooth it out again.

Kneel down on one knee with the other foot flat on the floor. Now watch your foot as it crawls forward like a caterpillar: the toes stretch out, reach far ahead of themselves and grab the floor; then the foot crunches up and pulls the heel along which then gets firmly placed on the floor. Do this backwards and forwards, first with one foot, then with the other. Now see if you can do it with both feet at the same time (you will have to sit for this).

Hold a ball or a piece of clothing in your toes and roll backwards, keeping your legs stretched as straight as possible. Put the object on the floor behind your head. Now pick it up again and return it to the starting position.

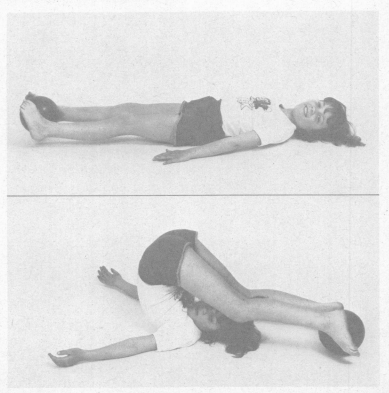

To tense the foot:

A good way to brace the transverse arch of the foot is to balance on a broomstick which is placed diagonally beneath the foot. Balancing on a broomstick is also good exercise for the whole foot.

A variation of the previous exercise can be done with a jump rope. Stretch the rope flat on the floor and try to balance on it. Place your heel squarely on the rope and walk along in such a way that the rope runs under the third toe. Now concentrate, and try to do the same thing with your eyes closed.

This exercise is intended to tense all the muscles in the foot. Sit on a chair or stool, place your heels firmly on the ground or floor, curl your toes in and turn your feet toward each other so that the soles are looking at each other while the heels remain firmly fixed on the ground.

This next exercise is a type of tug-of-war with your feet. Two partners stand shoulder to shoulder but face in opposite directions and spread their legs apart. Each child hooks his right foot against the inner edge of his partner's right foot. Who can pull the other's foot out from under him? Repeat with the left foot.

Children can be a great help in the garden. For instance, they can help mow the lawn by vigorously pulling the grass out with their toes. Who can collect the biggest pile of grass in the shortest time?

Adequate exercising of the foot and toes is very important for maintaining good posture; one should, for instance, walk and stand as often as possible on tiptoes. This can be done while brushing your teeth as well as while doing daily chores or helping in the kitchen. Parents would also do well not to be over-eager in fetching things for the smaller children which might be just beyond their reach. Children should make a good effort at fetching the object themselves first by getting on their toes and fully extending their backs. Only when it is impossible should the parent intervene and fetch the object.

Warning: *"Walking on the outer rim of your feet" used to be recommended as a good exercise for the foot. This has proven to be actually injurious, because it overextends the outer ligaments and the outer articular joint of the foot, and it does no good either for the arch of the foot or for its tension.*

Conquering obstacle courses is another source of fun for children. A circular obstacle course can be made by using a jump rope, a medicine ball, a broom, and books of various sizes. The course, with a little imagination can be expanded or varied at will to meet individual needs. The object of the game is to see who can master this course without falling off the broom stick, without toppling one of the books while jumping over the pile, and without stepping on the rope while criss-crossing over it. Other skills can also be incorporated, for instance, who can be the first to jump off the medicine ball after balancing on it with both feet for a short or specified length of time. One thing to remember with games like these is that there should also be winners, for this aspect increases the attraction.

In this next exercise, the child can pretend to be a tall stalk or reed on a sand dune. A number of participants (at least two) form a tight circle (or face each other in the case of two people). The child stands in the middle and is gently pushed back and forth between the others like a reed in the wind. The "reed" has to keep his body well tensed and stretched, and he has to use his feet to catch and balance out the respective body position. In doing so, he is not allowed to budge from the spot and cannot move his feet to regain his balance. Every child should have a chance at being in the middle.

Exercises for Knock-Knees

Goal: *to strengthen the muscles running along the inner side of the leg which help stabilize the knee, to foster an outward displacement of the stress points and thus of the growth pressure in the knee joint, and to strengthen the rump muscles.*

Many children have seen this "trick" performed in the circus. Have the child lie on his back on the floor or grass and hold a ball tightly with his feet. The object is to throw the ball to his partner. To do this he has to draw his legs slowly toward his body in preparation for the throw, then stretch them out quickly and vigorously so that the ball flies away.

In this exercise the child bends his knees a bit and holds a ball or another object between his knees. His partner then tries to pull the ball away. After doing this a few times, the child holds the ball again, but this time between his feet. The object of this exercise is to bend down and straighten up again without letting go of the ball.

In this exercise the child can pretend his legs are the blades of a scissors. All participants sit on the floor with their legs stretched out in front of them and their hands behind their back, supporting their body weight. As the "scissors" open, the child draws his feet toward his body and presses the soles of his feet together. Draw the feet as close as possible to the body, and then close the blades by stretching the legs out again.

This is another "trick" the child has frequently seen in the circus. Place two jump ropes side by side on the floor or ground to form a long line. The first "tightrope" skill is to walk alongside the line in such a way that the inner edge of each foot touches the rope. Then, with stalking steps and knees held high, the child should balance on the rope and finally crisscross over the rope with every step, so that the rope runs along between the outer edges of each foot.

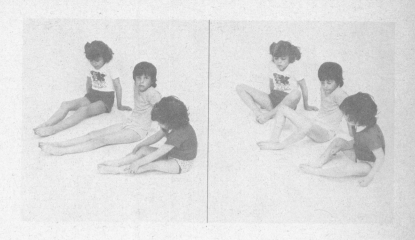

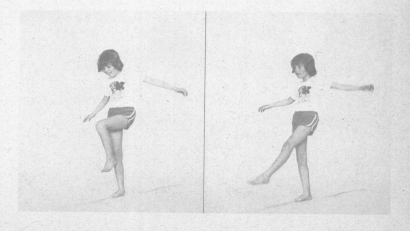

To do the "bicycle", two partners lie on the floor facing each other and press the soles of their feet together. The object is to continue peddling for as long as possible without losing the foot contact and being careful not to lessen the resistance. Then stretch the legs out straight and spread them wide apart in the air and close them again, all the time being careful to keep the soles pressed together.

Two children lie head to head in the supine position and pass a ball or a piece of clothing they are holding in their feet over their heads to the other partner. The partner then places the object on the floor, picks it up again with his feet, and passes it back to his partner.

Two partners sit facing each other in the spread leg position. The feet of one of the partners are placed outside the feet of the other. Now each partner takes turns in trying to open or close the "scissors" by tensing his or her leg muscles.

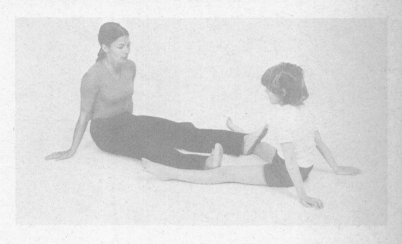

In this exercise, the child pretends to be a hardworking winch. The child lies on his stomach over a thickly upholstered sofa armrest or on a table that has been covered with a blanket, so that both legs hang down over the edge. If he now holds a teddy bear or a doll between his feet, his legs become a proper "winch" which closely raises and lowers itself. The winch will probably work better if the partner "cranks" on the edge of the table or sofa and the "winch" makes a loud rattle.

This next exercise should be done outside and can be done with a group. Mark out a course and have each child jump around it while keeping one foot on top of the other. The next time around the course, the child should hop with his feet crossed one over the other. Jumping like a frog is also good for competition: the child crouches deeply with widespread legs and then hops up and straightens his legs at the same time.

Exercises for Overextension of the Knee

This exercise is a variation of the well-known knee-bend. The child should keep his upper torso very straight and slowly sink into a semi-knee-bend and then straighten up again, equally as slowly. His partner, in this case preferably an adult, kneels behind the child and uses his or her hands to give a slight resistance in the hollow of the child's knee so that it cannot bend backwards. In this way the child gradually develops a feeling for the proper knee position.

8. Exercises for Preschoolers

Exercises For Curvature Of The Hip (Hip Kyphosis)

Goal: *to stretch the muscles that run along the back side of the leg, and strengthen the deep hip flexor muscles in order to pull the pelvis forward and thus attain a complete extension of the back in the upright sitting position.*

To stretch the hip and leg muscles:

Start this exercise in the upright sitting position. Each participant rolls a ball along his outstretched legs from his feet to his waist without bending his knees even a little bit, then around the back of his body and down his legs again.

This next exercise starts out from a crawling position on all fours. Each participant stretches his legs straight out and "walks" closer and closer to his hands that are resting on the floor supporting the body weight. The object is not to bend the knees. Once you have reached your hands, walk back again, again being careful not to bend your knees. Who can come closest to his hands?

Another exercise reminiscent of the circus is the elephant walk. Each participant marches several times around the table in elephant steps, on all fours, that is, but keeping the knees stretched straight.

To strengthen the hip flexor:

This exercise can be done on a flat surface like a floor or a lawn. The child lies in the supine position and his partner, best an adult, who is kneeling in front of him holds the shin of his bended legs so that the child can push himself vigorously away. To do this he has to stretch his legs and, what is more important for our purposes, he has to pull himself back again.

The supine position is again the starting point for this exercise. Each participant pulls his legs as tightly as possible toward his stomach without using his hands. He has to keep them in this position while his partner, preferably an adult, tries to pull them away again. Once in this position, the child can even try to raise himself to a sitting position and lie back again without letting his legs stretch out straight.

Each parent holds one end of an athletic wand high enough off the floor to allow the child to dangle from it. A sense of excitement can be introduced if all participants pretend that the living room floor has turned into an ocean and the child may not get his feet wet, not even when his parents let the horizontal bar sink lower and lower toward the floor.

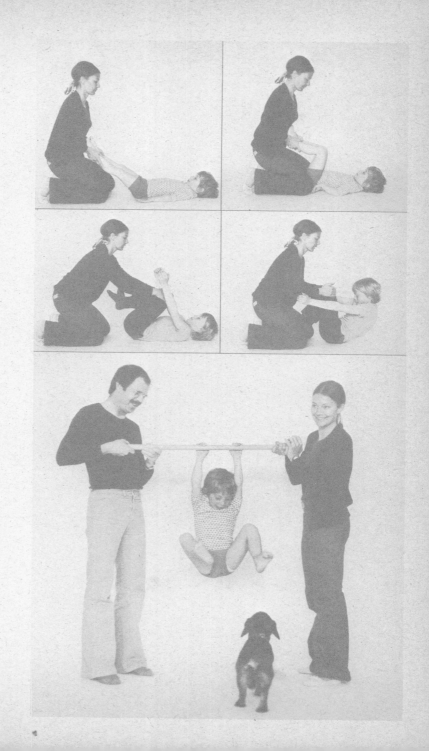

To extend the back:

In this exercise the child's father becomes the "winch". He lies in the supine position and bends his legs far enough that the child can rest on his stomach on top of the soles of his father's feet. Father and child hold hands, and the winch raises when the father slowly stretches his legs. The child should float freely and fully stretch his back and legs.

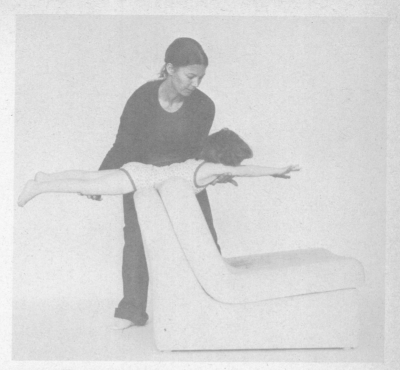

This time the child pretends to fly like a bird. He lies on his stomach over an upholstered chair or sofa back (or armrest) in such a way that he can stretch his legs out behind him. He then holds them in this outstretched position for a few seconds at a time, and then lets them sink down again. When finished, a good way to end is to somersault into the chair, but the child should be reminded to tuck his head in before rolling forward.

Another variation of essentially the same exercise lets the child fly on a carousel. The child lies on his stomach over a piano stool or over an adjustable stool and lets his arms and legs dangle down. At a signal the child slowly raises and stretches them to full length. If the stool should revolve, he or she is flying on a carousel.

To do this exercise, all participants should lie in the supine position. With the knees bent and the heels drawn as close as possible to the torso, lift your bottom off the floor as high as you can, hold it there for a few seconds, and then sink down again.

Exercises For Lumbar Flexure And Protruding Stomach

Goal: *to strengthen the back extensor muscles and the stomach muscles. Also to acquire a secure feeling of body posture.*

To strengthen the stomach muscles:

The adult partner, the child's mother, for example, sits on a chair or a stool with her legs stretched apart. The child sits on her lap and slowly lets himself fall backward until he is lying outstretched on her legs. In the same way the child then slowly rolls back to the sitting position, starting the movement by raising his head. If the child can do it alone, he should stretch out his arms while returning to the sitting position; small chidren, however, should hold on to their mother's hands.

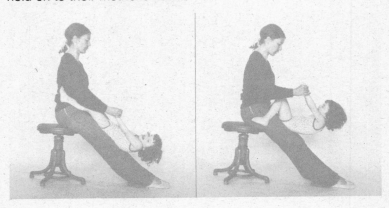

In this exercise the child sits with his legs stretched out in front of him and places a ball on top of his feet. If the child slowly raises his legs, the ball will roll toward his stomach, if he lowers them again, it will roll back toward his feet. If the ball doesn't do this by itself, the child can raise his bottom a little so that the ball is again brought into motion.

In this exercise the child begins in the supine position. His outstretched legs form a winch which slowly lifts the teddy bear he is holding in his feet high in the air and slowly lowers him again. The object is not to drop the teddy bear.

When the child is lying in the supine position on the floor and a toy airplane flies in lower over his tummy, the child has to draw in his stomach and hold the tension long enough for the airplane to return and finally land in the middle of "tummy" airport.

The following exercises are of help in straightening out a lumbar flexure and extending the back:

In this exercise the child can pretend to be a rocking horse. He starts out in the supine position and uses his hands to hold his bended knees tightly over his stomach. This will round the child's back and enable him to rock back and forth.

This time everybody is a bridge. Starting from a crawling position, arch your backs high enough so that a big ball can roll through or one of the smaller children can crawl underneath you.

Three-year-olds can already do a somersault on a flat gymnastic mat, but they have to make sure they tuck their heads in first before they start the forward roll.

The child lies on his stomach and rolls a ball to his mother who is kneeling in front of him. To do so he holds the ball in his bended arms and pushes it away.

This exercise is called the inchworm. The participants inch their way forward on their stomachs by stretching their arms out as far as they go and wiggling forward like a snake.

This next exercise involves a bit of competition. Who can creep under the table and chairs in a completely flat position? The tunnel can also be made out of loose chair or sofa cushions propped up against each other.

Every child likes to ride on his father's shoulders. While pretending to be a "wild horse", Father should hold the child securely by the legs so that the obstinate "horse" can make all kinds of stubborn movements. The rider has to use his arms to keep his balance while the "horse" leans forward, twists and turns to either side, and might even make irregular hops now and then. Just remember never to bend backwards, for this is not good for the horse or the rider.

Little ones like to climb and balance. They can climb up on their fathers and try out all sorts of tricks together; for instance, Father can kneel down on one knee, support himself with the other leg and stretch his arms wide apart so that he soon resembles a climbing tree. There are also other climbing exercises of this type: in one the father stands erect, places one foot in front of him to brace his weight and holds the child by the hands while the child climbs up his leg to his father's shoulders.

The child's father can also lie on his back and bend his legs so that the child can ride on his knees. This is especially suitable for morning exercises before getting up, for it only adds to the fun if the rider should actually fall off the horse.

The child can also balance and ride on his father's back while he crawls along on all fours. The child has to press his legs tightly around the "horse's" waist so that the horse can arch his back and then sink down again. At times the child may even be allowed to stand on the "horse's" back if his mother holds the child securely by the hands. In order to be a successful circus rider, the child has to keep his balance, and this requires the coordinated interplay of the feet, the legs, the pelvis and the back extensor muscles.

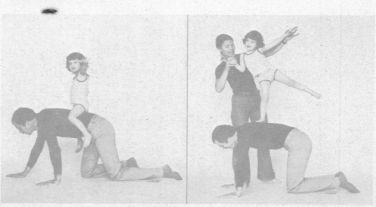

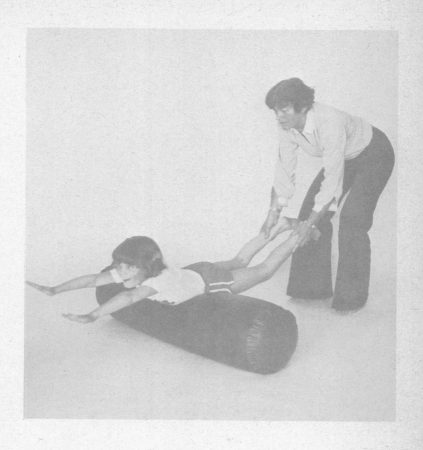

There are also other balancing acts a child can do which will also help his posture. Let the small "acrobat" balance himself on top of an inflatable training bag or a large ball (on his stomach). Hold the child's legs so that the arms are free and the child can stretch his back.

Four- and five-year-olds approach these acrobatic exercises involving a ball or a training bag with a great deal of enthusiasm and imagination. One should not be too anxious about this, but should remove all sharp or pointed objects from the room before the exercises begin.

Parents or adults can help the child in this balancing exercise too. The adult should sit on the bag behind the child and hold him by the legs. When the bag starts to roll back and forth a little, the child has to try to keep his balance. This is a good opportunity for the child to imagine he or she is a ship captain weathering a storm.

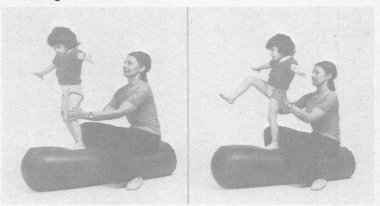

Exercises For Hunchback And Funnel Chest

Goal: *Small children very rarely have a real hunchback. This can be demonstrated by lifting the child by the legs or laying him flat on his stomach, for in these positions his back extends itself and becomes flat and straight. The posture problem here, the rounded shoulders, has not yet become fixed.*

Only with bigger children can a rounded back be so seriously pronounced that it can also be reflected in the position of the stomach. If the round shoulders become a habit, they move more and more forward so that the muscles which border on the front arm pits become increasingly shortened. In order to avoid this, these muscles have to be extended, the spine has to be kept flexible, and the muscles that extend the back have to be strengthened, too, for they provide the front support of the torso, they support the internal organs and see to it that their weight does not pull the back into a lumbar flexure.

All hanging exercises stretch the forward armpit muscles, whether they be done on rings, on a cross bar or on an athletic wand which the parents hold high between them. The child's hands should grasp the bar at shoulder width so that the chest and the shoulder girdle are not pressed too tightly together. Hanging with the head downward, if the father for instance dangles the child by his legs, is another good way to stretch the back, especially when the child can touch the floor with his hands. In this way his arms are used as supports and the back is extended to its full length.

103

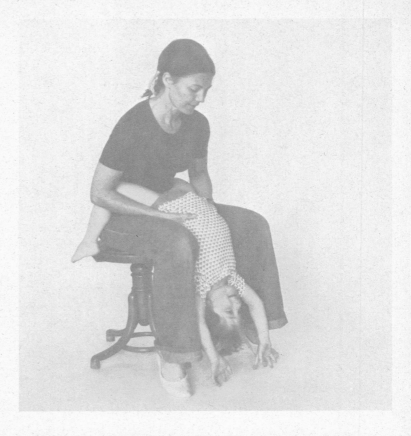

A good way to help smaller children stretch their backs and mainly their thoracic vertebrae completely is to let them bend backwards on their mother's lap and try to touch the floor with their hands. When straightening up again, remember to tell them to raise their head first, then the rest of their body. It is best to hold smaller children under the arms to thus support the torso; bigger children can pull themselves up by holding onto the adult's hands.

All ball games that are played while lying one one's stomach stretch the back. The child can roll or throw the ball from this position. Both of these movements use the arms for throwing, not supporting the body, and the back thus has to hold itself up.

In this exercise, which is also aimed at stretching the back, the child can pretend to be a circus performer again. The object is to carry a weight on his head while he walks straight and tall for a certain distance. The "weight" need consist of nothing more than a filled hot water bottle, a small bag of sand, or a book.

An overextension of the thoracic vertebrae is a very good way to fight funnel chest. Place a solid pillow or the training bag under the shoulder blades while the child is lying on his back. The child then stretches his arms upward and over his head, inhales deeply and makes himself as long as possible, then exhales smoothly.

Another way to help a child stretch his back is to let him lie on his stomach across a training bag, large ball, or across his mother's lap and stretch his arms out wide like wings. This, too, will stretch his back completely.

Exercises for Protruding Shoulder Blades

Goal: *to stretch the front chest muscles that pull the shoulders forward, to strengthen the shoulder blade muscles, especially the ones that run between the shoulder blades and the spine.*

All push-up and bracing exercises are suitable for flattening out protruding shoulder blades.

In this exercise the child lies flat on his stomach and crawls on his elbows and knees under tables and chairs, under a broomstick that is held rather close to the floor, or under two sofa cushions that have been propped up against each other to form a tunnel.

Almost every child knows how to do the "wheelbarrow". One child walks on his hands while his partner holds his legs up. The object is to see who can keep his back the straightest and longest and not stick his bottom in the air. With smaller children, the adult should support the child's stomach by placing the training bag underneath it.

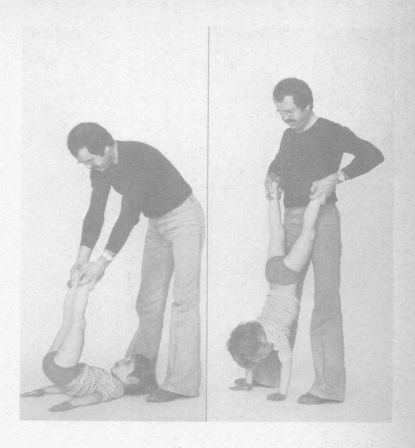

This next circus trick involves a bit of skill and is all the more satisfying because of it. The idea is to roll backwards and end up in a handstand. The child lies on his back, his father stands behind the child's head and catches the child's legs which are stretched up toward him. The child then places his hands palm down near his head and while the father lifts the legs, the child presses down hard with his arms so that he rolls backwards and ends up in a handstand. Stay in this position for a few moments, and then walk away like a wheelbarrow.

107

Little children can pretend to be a bridge in this exercise. First they sit on the floor, place their hands palm down behind them for support, place the soles of their feet flat on the floor, and lift their bottom as high as possible.

This next exercise calls for "windmills" and "bees". The child can pretend to be a windmill by swinging first one arm and then the other in large circles at his side. From this position he can magically turn into a "bee", for they make very small circles with both arms at the same time. One thing to remember, though, is that a bee's wings are held at shoulder height.

This time the child lies on his stomach and crawls forward on his elbows like a crab, trying to catch his mother who is running away from him, quite frightened.

This next one is another variation of the rocking horse exercise. The child lies on his stomach and holds his feet with his hands. His mother or father may need to help him hold his feet tightly as he lifts his body from the floor. They should also make sure that the child does not make a lumbar flexure and that only the front chest wall is extended. When done properly, this exercise is also good for children with funnel chest.

A similar exercise can be done outside. All participants swing their arms in very wide arcs as they run around in a circle in the fresh air. Remember to swing only the outside arm, change directions, and then swing the other outside arm.

Round shoulders and protruding shoulder blades occur together so frequently that the exercises for both of these conditions are interchangeable.

9. Exercises for School Children

Exercises For Curvature Of The Hip (Hip Kyphosis)

Goal: *to extend the muscles running along the back of the leg and to strengthen the deep hip flexor in order to pull the pelvis forward and thus attain a complete extension of the back.*

This first exercise involves a little competition with the other children. All participants sit on the floor, draw their knees up to their chin, grasp their ankles with their hands. The object is to stretch out the legs as far as possible without letting the ankles go.

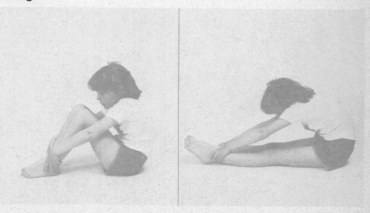

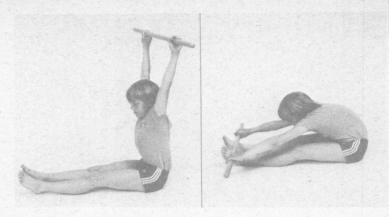

Again starting from the upright sitting position with their backs straight and tall, the children use both hands to hold the athletic wand. First they raise it high over their heads, bend far forward, and place the wand on their feet without bending their knees.

This next exercise is a familiar one. All participants sit on the floor in the spread leg position, raise their arms over their heads, stretch their backs tall and straight and bend first toward the right, then between the legs, and then toward the left leg. Remember to keep the pelvis pulled forward and to straighten up between each change of direction.

Two or more participants can make a race out of this exercise. All fold their hands behind their heads, stretch their backs and inch forward by twisting first the right hip and pelvis, then the left hip and pelvis. Remember to keep the back fully extended at all times.

This next exercise will require a little practice. The child starts out by sitting on his heels and then crouching down, resting his head on his knees, letting his arms hang next to this body, and turning the soles of his feet upward. From this starting positin, the child raises his head and stretches his arms out behind him so that they pull the shoulders back. It is important to stretch the hands right to the fingertips, keeping the palms turned inward. The child's back will form a tensed flat surface, and his bottom will be slightly raised up. Depending upon his strength, the child should hold this position for a few seconds, and then fall easily back into the starting position. Try it again.

This time the child starts in the upright sitting position with his arms propped up behind his back and the tips of his fingers pointing toward his feet. The first part of the exercise is to let the small of the back droop easily toward the floor so that it makes a curve. The second part is to pull the pelvis slowly forward until it is perpendicular and the upper part of the body is stretched tall. The child should hold this position without the help of his hands for a few seconds. Then relax, and return to the starting position. When repeated frequently enough, this exercise transmits a secure feeling for the position of the pelvis and the straightening of the back, especially when it is done in front of a mirror and the child can visibly "check" his movements.

Exercises For Lumbar Flexure And Protruding Stomach

Goal: *to strengthen the trunk musculature, both the back extensors as well as the stomach muscles, thus acquiring a feeling for the position of the pelvis and a stretched torso posture. This also requires a good flexibility of the spine in order to be able to correct the posture.*

To strengthen the stomach muscles:

All participants start out in the supine position on the floor, spread their legs apart and stretch their arms behind their heads, keeping them about shoulder-width apart. Now lift the right leg and touch the right foot with the left hand. Then reverse the procedure and practice with the left foot and the right hand.

In this exercise, the child lies on his back and places his legs on the thighs of his mother who is kneeling in front of him. The child's hands crawl along his legs until they reach his feet and then crawl slowly back again. As he does this, the child slowly lifts himself from the supine position with his arms outstretched and ends up in a sitting position. He then returns to the original supine position again. To vary the exercise, the child should first do this with his legs stretched out straight, and then try it with his legs bent at the knee.

School children enjoy the bicycle exercise as much as adults do. While lying in the supine position on the floor, all participants "pedal" a bicycle in the air with strong pushes, making sure that their feet are working, too, by pointing their toes. The pedal movement will slowly flatten out until it is quite close to the floor and it should then gradually climb back up again. To end the exercise, the participants should draw their legs together to form a "candle" in the air and slowly lower them to the floor, keeping the small of their back flat on the floor.

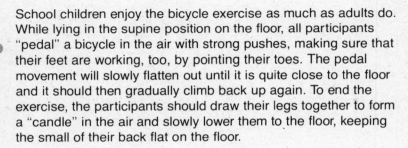

Starting in the supine position with the knees bent and the soles of the feet resting on the floor, each participant in this exercise alternates extending the right and then the left leg straight out and drawing it back toward the body. The whole movement should be kept as close as possible to the surface of the floor.

114

This next exercise will also require a bit of practice to perfect it. Start out in the upright sitting position on the floor, draw the knees up to the chin, and stretch out both feet at the same time, keeping the arms stretched out at shoulder level in front of the body. Hold the stretched position for a short time, and then return to the starting position.

The child will need his mother's help for this one. He starts out in the upright sitting position with his legs spread apart and his hands folded behind his head. While his mother holds his ankles, the child bends backwards, first to the left and then to the right, touching the floor with his elbow. After each touch the child straightens up completely and stretches his back before bending to the other side.

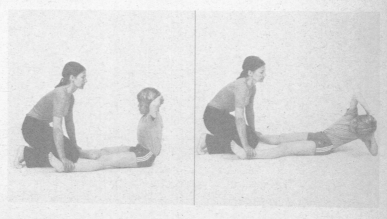

The following exercises are intended to help straighten a lumbar flexure, to loosen the lumbar region of the spine and to stretch the back:

Starting from the supine position with their feet flat on the floor and their knees bent, the participants press the small of their backs strongly against the floor, hold this position for a few seconds and then relax. During the relaxing phase one should bend the small of the back upwards in order to feel the difference in the position of the pelvis. Then repeat pressing the small of the back against the floor. Finally, stretch the legs straight out and again press the small of the back against the floor, hold the position for a few seconds, and then relax.

For this exercise, the participants lie on the floor with their knees drawn close to their stomach, their arms stretched straight out from their body at shoulder height. The object is to alternate twisting the body to the left and then to the right, keeping the legs in the bent position and deliberately making the small of the back round at each turn to the side. When lying flat on their backs again, or, in other words, in the middle position, all should draw their legs tightly against their stomachs. Their arms and shoulders should remain flat on the floor.

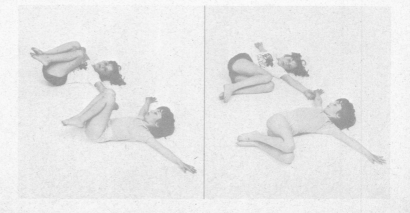

This exercise is similar to the previous one. The child lies on his back on the floor with his legs stretched out straight and his arms stretched out at shoulder height. While his mother holds his upper arms on the floor, the child moves his legs like windshield wipers, first to the left, then to the right. This exercise stretches the front chest muscles.

This one also starts in the supine position. All participants pull their heels as close as possible to their bottoms, then raise their bottoms from the floor without actually standing on their toes. The object is to alternate stretching first the right, then the left leg straight out while keeping both thighs at the same height.

The basic components of this exercise are familiar to all by now. The participants sit crossed legged on the floor, fold their hands behind their heads, and touch their left knee with their right elbow, and vice versa. It is important to remember to straighten up completely before changing direction.

Sitting cross-legged on the floor, all participants use both hands to hold an athletic wand over their heads. They then bend far enough forward for the wand to touch the floor. The second part involves straightening up and pulling the wand as far as possible behind the shoulder blades. This exercise should be repeated several times.

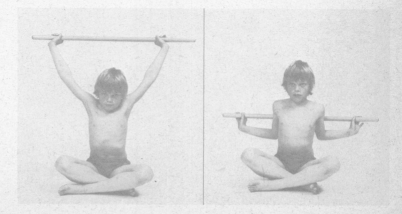

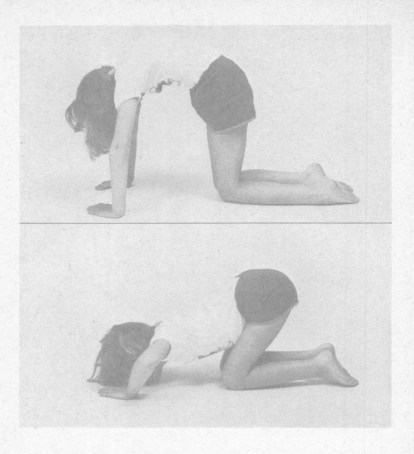

In this exercise, everyone leans on all fours and alternately arches his back upwards and lets it sink down again, making sure that the head and neck also go along with the movements. After repeating this several times, each person should shift his weight while arching his back so far behind that he is almost sitting on his heels. The object here is to let the back sink down again and push the torso as flat as possible forward between the bent elbows. The last phase of the exercise is done by rounding the back again in a smooth movement and straightening the arms out, thus returning to the starting position.

Kneeling on all fours, each participant draws one knee far enough forward so that it touches his or her nose. The next part is to raise the head, stretch the neck and stretch this same leg backward while at the same time raising it upwards.

To do this exercise properly, the child gets on all fours and raises the left arm and the right leg at the same time, stretches them out, returns to the original position and repeats the movement with the other arm and leg. The child should follow his raised hand with his eyes and concentrate on his palm. At the same time, he should remember to point the toe of the outstretched leg outward.

It takes two people to do this exercise. The child sits on his heels with his mother facing him at his side. In order for the child to get a sense of the position of his pelvis and the connection between a lumbar flexure and a protruding stomach, his mother places her one hand on his stomach, the other on the small of his back. She then tells the child to round out the small of his back by pushing it toward the back and then to pull it back in again, thus making a lumbar flexure followed by making the stomach protrude. While the child does this, the mother provides slight resistance with her hands.

Exercises for Hunchback and Funnel Chest

Goal: *to extend the front chest muscles which are pulling the shoulders forward because of their increasing constriction. The spine should also be made more flexible and the muscles that stretch the back strengthened in order to hold the torso upright and straight.*

All exercises that involve hanging by the arms stretch the front armpit muscles, whether they be done on rings, on a horizontal bar, or on a chin bar fixed in the door frame. While doing these kinds of exercises, the child should keep about a shoulder's width between his hands so that the shoulder girdle and the chest do not squeeze together. Bigger children can also hang by the hollows of their knees, first getting a good hold and then hanging freely and loosely in an outstretched position with their head downward.

The front chest muscles can also be stretched when the child lies on his back over a rolled-up blanket or over a training bag and moves a wand over and behind his head, places it on the floor, and picks it up again. Again, he should remember to keep at least a shoulder's width between his hands. Also appropriate for stretching the armpits is simply lying down and stretching the body as far as possible, including the fingers and toes. This exercise counteracts funnel chest.

This next exercise is primarily a stretching one. All participants lie on their stomachs and make themselves as tall as possible by stretching from their finger tips at the end of outstretched arms to their toes at the end of outstretched legs. The next part requires that everyone lift their arms and legs a bit from the floor, hold the position for a short time and then relax. Alternate stretching the right and then the left side by stretching out the right arm and leg at the same time, then the left arm and leg at the same time. Finally, raise both arm and leg of the same side, hold, and then relax. Follow this by doing the same thing on the diagonal, in otherwords, lift the right arm and the left leg, hold, relax, and then

Another stretching exercise also starts out in the stomach position. The child folds his hands behind his head and raises the upper part of his body slightly, holds the position, and then returns to the original position. Remind him not to snap his head, but to stretch his neck instead.

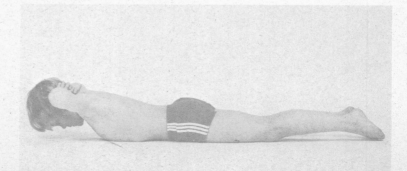

This time the child squats down on his heels with his arms held loosely next to his body. The first part is to stretch and tense the torso; this will straighten the child's back and raise his bottom up a little. Remind him to stretch his arms to the finger tips with the palms facing his body. He should hold this position for a few seconds and then sink slowly back to the starting position.

This exercise is similar to the one above, except that this time it is done with the hands folded behind the head. Each participant slowly straightens out the torso with his elbows stretched back until he gets into the heel sitting position. If it will help, a partner can hold the child's ankles to give him a bit more support.

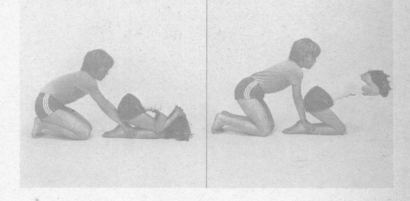

This next exercise requires a partner. The child sits crossed legged on the floor and folds his hands behind his head. His partner kneels behind the child and provides a slight resistance against the child's elbows which will strengthen the extension of the child's back. The child then lets the upper part of his body fall forward, which will also pull him forward, then he straightens himself up again, equally as slowly, against the slight resistance of his partner. He returns to the full stretch position.

This exercise can be especially useful as a breathing exercise. The child inhales while stretching the trunk of his body against the resistance and exhales while he falls forward, making a "fff..." sound as the air passes through his slightly open lips. After a few tries, alternate the procedure: inhale while rounding the back, then straighten up, exhale and stretch, without letting the torso sink back into itself at the end of the exhale. The latter sequence is especially important for the overall posture, for one should keep the back extended not only while standing, but while walking and sitting as well, in order not to sink into a slouch with every exhale.

Another way to attain an extended body posture is to carry an object on your head such as a medicine ball, a small bag of sand, or a book. Balancing also requires stretching the body. This is why parents should make use of every appropriate tree branch they can find while taking a walk with their child. It is also important to vary the exercise: the child should try to keep his balance on the branch or fallen trunk with his hands folded behind his head and with his elbows drawn back or with his arms outstretched, and if he is careful enough, he should even try to go backwards on a broad tree limb.

The exercises described above are also useful in counteracting protruding shoulder blades and they are thus interchangeable with the following ones.

124

Exercises For Protruding Shoulder Blades

Goal: *to extend the front chest muscles that pull the shoulders forward, strengthen the shoulder blade musculature, especially the muscles that run between the shoulder blades and the spine.*

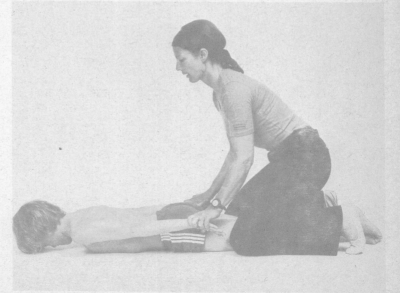

In order to crawl under a chair, under a broomstick that is held close to the floor, or under his mother who is kneeling on all fours, the child has to make himself very flat. In some cases he must even lie flat on his stomach and only "crawl" forward on his propped up elbows or inch his way along like a snake when the passage is almost too low to get through.

In this exercise the child lies on his stomach on the floor with his mother kneeling over him. She grasps his hands, which he holds at his sides, and pulls his shoulders back so that the upper part of the child's body rises slightly from the floor. The child then braces himself tightly in his mother's hands which provide the needed resistance, holds the tensed position for a few seconds, and then returns to the starting position.

This time the child starts off in the supine position. His arms lie slightly bent next to his body. When he tenses the trunk of his body, his arms will prop the shoulder girdle off the floor. He should hold this tensed position for a few seconds and then return to the relaxed starting position. Be sure he does not make a lumbar flexure.

Bigger children, too, can make a bridge by propping up their hands and feet while lying in a supine position and then raising themselves from the floor. To counterbalance this, the child should return to the supine position and then roll backwards until his feet touch the floor behind his head. This exercise should be repeated a few times.

A push-up requires strong arms to hold the body up. It is easier for a child if he can put his legs on a low chair or on a training bag and then prop himself off the floor with his arms. To do so he should first bend and then stretch his arms. Another phase can be added by alternately raising first the right and then the left arm stretched out in front of the body.

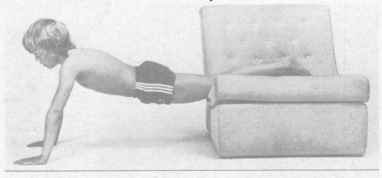

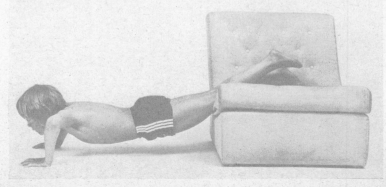

Starting from the all fours position, this time the child pushes one leg out behind him without lifting it off the floor and supports himself with his toes on the floor behind the kneeling leg. With the arm of the same side he now draws a large circle in the air: forward and high and in a wide arc over, above, behind and back again. The object is to let the arm stretch out well, and follow the hand with his eyes. It is very important to do this exercise slowly, so that each side of the body can stretch and tense. Change sides and repeat.

This next exercise also starts out in the all fours position. The child bends his arms so that the elbows are pointing outwards. In this flattened posture he then crawls forward with small steps.

To do this exercise, the child should sit on his heels and crouch down as low as possible. He first clasps his hands behind his back and pulls his shoulders back. This stretches his back and tenses it horizontally. The child should hold this tension for a short time, then sink down again and relax.

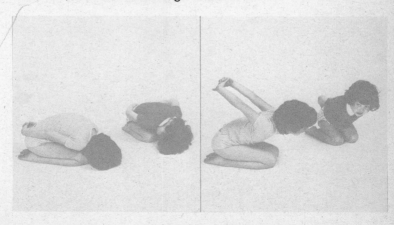

The starting position is once again on all fours. All participants should sit back until they are in the heel sitting position. While doing so, each one pushes his outstretched arms far enough forward so that his upper body hangs loose and is only supported by his hands that are resting on the floor. The bottom is raised a bit from the heels while doing this. The first part is to bounce a few times, then raise first the right and then the left hand and look at the palms. Then, in this same position, each person now crawls forward on his knees with small steps, whereby each step with the knee is preceded by stretching out the respective hand as far forward as possible. The knee then follows the hand and produces the step.

In this exercise, two children of approximately the same size sit cross-legged and back to back, put their hands on their hips, stretch their backs and press their flexed arms against each other. They hold this position for a while and then relax. They then fold their hands behind their heads and press their elbows against each other. The tension that results from this should not be very great.

This final exercise should be done outside with a partner. The child spreads his legs apart and stretches his arms over his head. The object is for the child to throw a ball high and behind his head. The partner, who is standing behind him, should catch the ball and roll it back to the child between his legs. The child then picks it up again. The partner can also throw it back sideways so that the child has to twist to catch it.